CAMP PENDLETON

CAMP PENDLETON

The Historic Rancho Santa Margarita y Las Flores and the U.S. Marine Corps in Southern California

A Shared History

by Breanne Robertson, PhD

Marine Corps History Division
Quantico, VA
2017

Library of Congress Cataloging-in-Publication Data

Names: Robertson, Breanne, author.
Title: Camp Pendleton : the historic Rancho Santa Margarita y Las Flores and the U.S. Marine Corps in Southern California, a shared history / by Robertson Breanne.
Other titles: Historic Rancho Santa Margarita y Las Flores and the U.S. Marine Corps in Southern California, a shared history
Description: Quantico, VA : Marine Corps History Division, [2017] (pbk.)
Subjects: LCSH: Camp Pendleton (Calif.)—History. | United States. Marine Corps—History. | Rancho Santa Margarita (Calif.)—History. | Historic buildings—California—San Diego County. | United States. Marine Corps—Biography. | California, Southern—History, Military.
Classification: LCC VE23 .R5 2007 | DDC 359.9/670979498--dc23

DISCLAIMER
The views expressed in this publication are solely those of the author. They do not necessarily reflect the opinions of the organizations for which she works, Marine Corps University, the U.S. Marine Corps, the Department of the Navy, or the U.S. government. The information contained in this book was accurate at the time of printing.

Published by
Marine Corps History Division
2044 Broadway Drive
Quantico, VA 22134
www.usmcu.edu/historydivision

1st Printing, 2017

CONTENTS

Acknowledgments VII

Introduction 1

The Mexican War, 1846–48 5

The Mexican Revolution of 1910 10

Major General Joseph H. Pendleton and the Origins of the Marines on the West Coast 13

World War II, 1941–45 19

Post-World War II Era, 1946–49 26

Korean War, 1950–53 28

Vietnam War, 1963–75 34

Post-Vietnam Era, 1976–present 37

Architectural Features of 1st Marine Division Headquarters, Building 1133 41

Architectural Features of the Santa Margarita Ranch House National Historic Site 53

Architectural Features of the Las Flores Adobe National Historic Landmark 73

APPENDICES

Appendix A:

Timeline 85

Appendix B:
U.S. Marine Corps General Residents, Santa Margarita Ranch House, 1947–2007 89
Appendix C:
Commanders of Marine Corps Activities, San Diego, California, 1914–42 91
Appendix D:
Base Commanding Generals, Camp Pendleton, 1942–present 94
Appendix E:
Base Sergeants Major, Camp Pendleton, 1956–present 97
Appendix F:
Commanding Generals of 1st Marine Division, Camp Pendleton, 1941–present 99

Further Reading 102

ACKNOWLEDGMENTS

This publication was created in cooperation with the 1st Marine Division and the History Museum Branch, Marine Corps Base Camp Pendleton in celebration of the 75th anniversary of the Marine Corps base at Camp Pendleton, dedicated on 25 September 1942. The author is especially grateful to Major General Daniel J. O'Donohue, Colonel Christopher S. Dowling, Lieutenant Colonel Maurice Brown, A. Faye Jonason, Robert S. Hellman, and Danielle M. Page at Camp Pendleton and to Dr. Charles P. Neimeyer, Douglas E. Nash Sr., Paul Westermeyer, Angela Anderson, and numerous other colleagues with the Marine Corps History Division at Marine Corps University for their support and encouragement in the making of this book.

INTRODUCTION

Southern California looms large in American culture as a land of sun and surf, but the region possesses a significant military presence as well. The importance of the Pacific coastline as a maritime gateway connecting the United States, Latin America, and Asia has made it an attractive outpost for trade and travel from the Spanish colonial era to the present day. An oft-contested terrain, Southern California changed hands no fewer than three times in the century preceding California's statehood. Since entering the Union, the region's mild climate and littoral perch, located as it is at a crossroads for both north-south and east-west hemispheres, has played a key role in U.S. military training and deployment, particularly for the U.S. Marine Corps.

Approximately 50 miles north of San Diego, the historic Rancho Santa Margarita y Las Flores (now known as the Santa Margarita Ranch House National Historic Site and Las Flores Adobe National Historic Landmark) has borne witness to the sweeping and often dramatic events that produced Marine Corps Base Camp Pendleton and the modern Marine Corps on the West Coast. Soon after the purchase of the expansive ranch land in 1942, Marines established a permanent training installation and returned to Southern California as part of the largest national mobilization effort for war since World War I. Nearby, the Santa Margarita Ranch House would serve as a primary residence for Marine Corps commanders for the next 60 years.

The immense construction efforts that transformed the sprawling ranch into one of the largest military installations on the West Coast included the Marine Division Command Post (designated by the U.S. Marine Corps as Building 1133). Modest both in design and scale, this

Oil painting of the Santa Margarita Ranch House by artist James Walker (1818–89), ca. 1885.
The Bancroft Library, University of California Berkeley

administrative building stands as a hallmark of civilian-military collaboration and wartime expediency in construction methods and materials. Today, it remains one of the few relatively unaltered temporary structures from that historical moment. Significantly, it also continues to operate as command headquarters for the 1st Marine Division.

The 1st Marine Division Headquarters building was one of the first structures to be completed at the new military base.
Photo by the author

The role of Camp Pendleton was essential to the development of an expeditionary Marine Corps both during and after the Second World War. Over the course of seven decades, the Marines assigned to the West Coast facility have demonstrated tremendous mobility and combat efficiency as the nation's foremost "force in

Main Gate at Marine Corps Base Camp Pendleton, ca. 1950.
Official U.S. Marine Corps photo

readiness." As the primary installation for training Marines for the Pacific theater during World War II, Camp Pendleton quickly expanded to accommodate the influx of recruits while simultaneously developing a rigorous training program in amphibious operations. During the Cold War, the military post again served as the gateway for Marines bound for combat in Asia. The wars in Korea and Vietnam each required expediency and flexibility on the part of the Marine Corps, which swiftly rebuilt its forces for deployment abroad. A continued emphasis on amphibious landings combined with naval and air support facilitated Marine combat and counterinsurgency efforts during these conflicts and secured the distinct mission and reputation of the Service for decades to come. In the post-9/11 era, Camp Pendleton continues to prepare Marines for rapid deployment to humanitarian and combat operations around the globe.

Since the founding of Camp Pendleton in 1942, base expansion, natural disasters, and decades of benign neglect have threatened to diminish the historic legacy of the Rancho Santa Margarita y Las Flores and the Marine Division Command Post. With the designation of Las Flores Adobe on the National Historic Landmark Registry and the Santa Margarita Ranch House complex on the National Register of Historic Places, the Marine Corps has engaged in preservation and restoration efforts to maintain these rich cultural resources. Moreover, the Marine Division Command Post (now known as 1st Marine Division Headquarters) has been deemed eligible for the National Register of Historic Places and may be added to the list in the future. Today, the Marine Corps proudly protects these heritage assets, preserving the rich history of the base and assuring their continuing place in military history for years to come.

THE MEXICAN WAR
1846–48

The history of Southern California and the U.S. Marine Corps has been intertwined since the mid-nineteenth century, when the young American republic, led by President James K. Polk, engaged in its first war conducted on foreign soil. Having been elected on a platform promising the "reannexation of Texas, reoccupation of Oregon," Polk set out to acquire the California territory from the Mexican government. Mexico had already refused an offer from the United States to buy the province a decade earlier; the country rebuffed Polk's purchase proposal again in 1845. Adopting an alternate plan, the president dispatched a 33-year-old Marine, First Lieutenant Archibald H. Gillespie, to courier secret messages to three U.S. officials on the West Coast. First Lieutenant Gillespie, disguised as a British whiskey salesman, arrived in California in April 1846. While the exact contents of his messages remain unknown, Gillespie instructed the recipients—American Consul Thomas O. Larkin, U.S. Navy Commodore John D. Sloat, and U.S. Army Captain John C. Frémont—to avoid acts of aggression yet also remain ready to strike in the event that war with Mexico should break out.

Following a skirmish in May 1846 that resulted in the loss of American lives along the Rio Grande River, the United States declared war against Mexico. Fighting along the disputed Texas-Mexico border soon spread to California, and as a majority of the U.S. Marine Corps' 63 officers and 1,200 enlisted personnel were on board Navy ships at the outbreak of war, the Marines assisted the U.S. Navy with maintaining a strict blockade of the Mexican coast. The Marine Corps also proved a flexible amphibious force whose landings at various ports and towns permitted a show of strength that made American

One of several campaign banners Nathanial Currier produced for the Democratic Party in 1844, this banner features portraits of the presidential and vice presidential candidates, James K. Polk and George M. Dallas.
Library of Congress Prints and Photographs Division, Washington, DC

numbers appear larger than they were. On 29 July 1846, a detachment of Marines and sailors from the sloop-of-war *Cyane* (1837) made landfall in the San Diego area. The squad occupied the town without opposition and raised the

The USS Cyane *Taking Possession of San Diego Old Town, July 1846* by Carlton Theodore Chapman.
Maritime Museum of San Diego

United States flag and Marine Corps banner over the Old Town Plaza.

While First Lieutenant Gillespie and a temporary garrison force remained in San Diego, the California Battalion of Mounted Riflemen (formerly known as the Bear Flag Army) began its march northward toward Los Angeles. On 13 August 1846, the battalion joined Commodore Robert F. Stockton's combined force, which included a Marine detachment from the USS *Congress* (1841), en route to their objective. The Mexican Army retreated upon the Americans' arrival, and Commodore Stockton drew up a plan for a territorial government and turned his attention toward enforcing the blockade of Mexico's west coast.

Portrait of Gen Andrés Pico, brother of the last governor of Mexican Los Angeles, Pio Pico, ca. 1847. As a leader of the resistance in Alta California during the Mexican War, Gen Pico worked to secure postwar protections for Californios under the 1847 Treaty of Cahuenga. He later became a citizen of the United States, commanded the California Militia, and held public office in the California State Assembly and California State Senate.
University of Southern California Libraries and California Historical Society

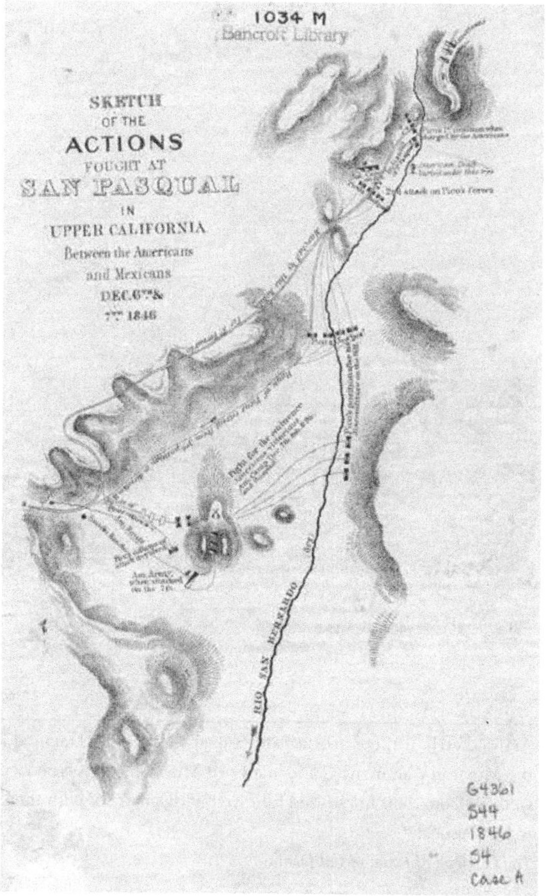

Sketch of the actions fought at San Pasqual in California between the Americans and Mexicans, 6–7 December 1846.
The Bancroft Library, University of California Berkeley

The battle for Los Angeles was far from over, however. In response to a local uprising, Marines and sailors conducted extensive training exercises on the San Diego shore in preparation for another march on Los Angeles. Marines were also sent among the reinforcements in early December 1846 to assist U.S. Army Brigadier General Stephen W. Kearny and his dragoons (cavalry units) in completing the final portion of their march from Santa Fe, New Mexico. At the Battle

Artist sketch of the Battle at San Pasqual showing U.S. armed forces fighting Californio lancers on horseback, ca. 1846–48. During the Mexican War, many California-born Mexicans preferred to use a lance as their primary weapon. They used a lasso to pull American soldiers from their horses and followed with an attack with the lance. This combination of weapons had a devastating effect at the Battle of San Pasqual.
The Oakland Museum of California

of San Pasqual, a joint force of dragoons under the command of Brigadier General Kearny and a detachment of Marines and sailors under the command of First Lieutenant Gillespie charged Californio lancers led by General Andrés Pico. The Americans sustained heavy casualties, including 21 deaths and injuries to both Kearny and Gillespie. Meanwhile in San Diego, Commodore Stockton formed a corps of mounted Marines. On 11 December, Navy Lieutenant Andrew Gray led a detachment of 80 Marines and 120 sailors as a reinforcement force to General Kearny's and First Lieutenant Gillespie's troops. The Californios, who had suffered fewer casualties and had surrounded the U.S. encampment during the engagement, were now outnumbered. General Pico and his troops retreated, allowing the American dragoons to depart San Pasqual unchallenged. Acting under orders from Commodore Stockton, General Kearny later commanded a mixed force of dragoons, sailors, volunteers, and Marines toward Los Angeles. The force engaged in several skirmishes along the march. On 10 January 1847, the Americans defeated a charge of Californio lancers and again occupied the city of Los Angeles.

The Battle of San Pasqual brought the U.S.-Mexican conflict literally to the doorstep of the Rancho Santa Margarita y Las Flores. While the proprietor of the ranch, Don Pio Pico, retreated to Mexico for the duration of the war, his brother-in-law John Forster provided 28 yoke of oxen from Pico's stock to support U.S. Captain Samuel J. Hensley's march toward Los Angeles. Meanwhile, Pico's brother Andrés remained loyal to the Californio cause. As commander of one of the three armies fighting U.S. forces, he used the ranch as a secure site to meet with his troops. According to lore, Californio officers Leonardo Cota and José Alipás, under Pico's command, planned the Battle of San Pasqual beneath a large sycamore tree near the Santa Margarita Ranch House.

In subsequent months, the Marines participated in several military actions alongside the U.S. Army and Navy in other parts of the province, while the Pacific Squadron successfully controlled thousands of miles of coastline along the west coasts of California and Mexico. Hostilities in the California theater of operations ceased about a month later. Performing a variety of services, ranging from amphibious landings to garrison duty to infantry charges, a total of 402 Marines served on the West Coast during the Mexican War.

The Treaty of Guadalupe Hidalgo formally ended the war in 1848, and Mexico ceded to the U.S. government a large tract of land that included California. Notably, it was General Andrés Pico, a prominent military leader in the government of Mexico, who signed the treaty at Cahuenga, now part of Hollywood, that ended the fighting between American and Californio forces.

THE MEXICAN REVOLUTION OF 1910

More than a half century passed before the Marine Corps resumed an active role in the history of Southern California. American dealings with Mexico again proved the catalyst, as anxieties about civil unrest south of the border prompted the United States to establish a visible military presence along its coast. Specifically, President William H. Taft determined to make a display of armed strength under the guise of training exercises to convey to Mexicans that U.S. nationals and property in that country must be respected. The U.S. Army moved units on both U.S. coasts, as did the Marine Corps. On the East Coast, the 1st Provisional Marine Brigade held training exercises at Guantánamo Bay, Cuba, while a provisional regiment, commanded by Colonel Charles A. Doyen, was hastily assembled at Mare Island Naval Shipyard, California, for "expeditionary service on the Pacific coast."[1]

The Mexican Revolution of 1910 underscored the advantages of San Diego as a strategic site for Marine Corps training and embarkation. Encompassing the 1st, 2d, and 3d Regiments, Doyen's newly established unit was designated the 4th Provisional Marine Regiment and transported to North Island, San Diego Bay, where it disembarked and established Camp Thomas in honor of Navy Rear Admiral Chauncey M. Thomas, commander in chief of the Pacific Fleet. When Mexican President Porfirio Díaz resigned a few months later and the turmoil subsided, part of the Marine Corps regiment at Camp

[1] *Annual Reports of the Navy Department for Fiscal Year 1911* (Washington, DC: Navy Department, 1912), 530.

Camp Thomas, 27 April 1911.
California State Military Department Historical Collection

Broadside view of USS *Jupiter* (AC 3) with the tug *Active* alongside in the Mare Island Channel, 16 October 1913.
Official U.S. Navy photo, #AC 3 002-10-13

Thomas disbanded, while the remaining officers and enlisted men returned to their regular duty stations in July 1911.

Continuing political and civil disorder in Mexico compelled the Marine Corps to return three years later. Following the coup and subsequent assassination of Mexican President Francisco Madero, the United States government refused to recognize General Victoriano Huerta as the legitimate head of state. Relations between the two nations deteriorated further in 1914 due to an incident at Tampico involving the alleged mistreatment of American naval personnel by Mexican officials and refusal of the Mexican government to apologize with a public gunfire salute to the American flag. Soon thereafter, President Woodrow Wilson retaliated by sending a combined force of Navy and Marines to the port city of Veracruz. Responding to intelligence that a European vessel with a cargo of arms and ammunition was en route to the Mexican coast, the commander in chief took preemptive action and ordered the Atlantic Fleet to prevent its delivery to the Mexican forces waiting ashore.

As U.S.-Mexico relations grew increasingly strained, Marine Corps officials considered it desirable to have a force ready to deploy for the west coast of Mexico. The 4th Marine Regiment was reactivated, and Colonel Joseph H. Pendleton assumed command of the expeditionary force. Its mission was to quell civil and political unrest in Mexico. Organized at Puget Sound, Washington, and Mare Island Naval Shipyard, California, in April 1914, the regiment embarked on board the USS *South Dakota* (ACR 9), USS *West Virginia* (ACR 5), and USS *Jupiter* (AC 3) for the Gulf of California, where it remained as a show of force. When stability was restored to the Mexican government the following July, the Marines returned to San Diego where they have remained ever since.

MAJOR GENERAL JOSEPH H. PENDLETON AND THE ORIGINS OF THE MARINES ON THE WEST COAST

Major General Joseph Henry Pendleton was a pioneering force in Marine Corps activities on the West Coast. An enthusiastic supporter of San Diego, Major General Pendleton acknowledged the strategic location of Southern California and advocated a stronger, more permanent Marine Corps presence in the region. His activism reflected an emerging Marine Corps doctrine of amphibious operations, which promoted the development of the "advanced base force" and began exercises as early as 1901. Arriving in California in 1914, following expeditionary service in Nicaragua, Santo Domingo, Guam, and the Philippines, then-Colonel Pendleton expressed an early commitment to establishing the Marine Corps as a force in readiness. He believed the mild climate and harbor of San Diego were especially well suited to this purpose.

In early July 1914, Colonel Pendleton's 4th Regiment disembarked in San Diego harbor from expeditionary duty in the Gulf of California, where it had been monitoring the Mexican Revolution as an American show of force. Whereas previous expeditionary units were disbanded and the officers and men returned to their regular duty stations, the Marine Corps maintained the 4th Regiment as the beginning of a permanent post in that city.[2] Under the command of Colonel Pendleton, the regiment established military quarters at North Island and dubbed the encampment Camp Howard in honor of the incumbent commander in chief of the Pacific Fleet, Rear Admiral Thomas B. Howard.

Colonel Pendleton considered North Is-

[2] A portion of the 4th Marines were stationed in San Francisco as well.

land unsatisfactory for a permanent Marine Corps installation, however. Two months after founding Camp Howard, the commander delivered a public address on the subject to a group of local citizens at the U.S. Grant Hotel in downtown San Diego. His lecture, "San Diego, An Ideal Location for a Permanent Marine Corps Base," expounded on the unsatisfactory conditions and inconvenient location of his regiment at Camp Howard as compared to the city of San Diego. Soon thereafter, he presented his argument to the Major General Commandant of the Marine Corps as well. In a letter to Brigadier General George Barnett, Colonel Pendleton noted the deplorable conditions at Camp Howard and introduced the idea of establishing a permanent Marine Corps base in San Diego. Located just 12 miles north of the Mexican border and containing an excellent harbor, the city readily presented itself as an expeditionary base, convenient to the Pacific approaches of Latin America, Hawaii, and the Far East.

Portrait of MajGen Joseph H. Pendleton (1860–1942), ca. 1917. When this photograph was taken, Pendleton held the rank of colonel.
Library of Congress Prints and Photographs Division, Washington, DC

Colonel Pendleton's 4th Regiment remained at Camp Howard until December, at which time the unit was assigned temporary duty at the Panama-California Exposition. In 1915, San Francisco and San Diego both hosted major celebrations to mark the completion of the Panama Canal, which had opened officially to maritime traffic the previous summer. The Marine Corps ordered the 1st Battalion of the 4th Regiment to the Panama-Pacific International Exposition in San Francisco, while the 2d Battalion maintained a model camp at Balboa Park as part of the Panama-California Exposition in San Diego. Among the highlights of the Marine Corps exhibit at the fairgrounds was a sham battle between cavalry and Marine forces on 6 April 1915.

Both the Navy and Army strengthened the

The 4th Marine Regiment, commanded by then Col Joseph H. Pendleton, at North Island, CA, in 1914.
Library of Congress Prints and Photographs Division, Washington, DC

Col Joseph H. Pendleton (far right), commanding officer of 4th Marine Regiment, and Capt Charles H. Lyman (third from right), adjutant of 4th Marine Regiment, at a display commemorating former President Theodore Roosevelt's expedition into the Amazon and River of Doubt during the Panama-California Exposition of 1915. Also pictured (left to right): unidentified Marine orderly; George Burnham, one of the organizers of the expeditions; G. Aubrey Davidson, prominent San Diego resident and president of the exposition; unknown; Ed Fletcher, politician and director of the exposition; President Theodore Roosevelt; unknown; and D. C. Collier, prominent San Diego resident and organizer of the exposition.
Private collection of Col Charles H. Lyman, USMC (Ret)

U.S. military presence during World War I and immediately afterward with permanent facilities in San Diego and surrounding areas. The Marine Corps also expanded in the region based on Pendleton's long-range vision. As early as 1914, Assistant Secretary of the Navy Franklin D. Roosevelt had made a trip to Southern California to inspect the San Diego area as a possible site for a Marine Corps "advanced base station" for the West Coast similar to the one being maintained on the East Coast at Philadelphia.

On 19 December 1914, Colonel Pendleton reported by telegram to Headquarters Marine Corps that the Marine Barracks San Diego had been established at Balboa Park. Despite this proclamation, the installation at San Diego remained in temporary status. No land had been acquired for a permanent station and no steps had been taken by the Navy Department to do so. Nevertheless, Pendleton persisted in advocating the unique suitability of San Diego as a location for a permanent expeditionary base.

In 1915, Assistant Secretary of the Navy Roosevelt made another visit to San Diego. He visited the exposition and again inspected sites that might serve as a future Marine Corps base. Following Roosevelt's return to Washington, Navy Department officials asked Commandant Barnett to inspect the San Diego locations under consideration and recommend the one he found most suitable. The Major General Commandant submitted his findings on 26 August 1915. Basing his opinion on location and purchase price, Commandant Barnett concluded that only one prospect—a tract on San Diego Bay—met the requirements for the current development and future growth of a Marine Corps installation.

Marines marching across Cabrillo Bridge during the grand opening festivities for the Panama-California Exposition, 1 January 1915.
California History Section Picture Catalog, California State Library

The need for a permanent West Coast base became increasingly apparent as internal conditions in Mexico prompted the Marine Corps to withdraw part of the 4th Regiment from exposition duty and send it on an expeditionary mission along the west coast of Mexico. In June, 300 Marines from the 25th, 26th, and 28th Companies of the 2d Battalion, 4th Regiment, departed the city on board the USS *Colorado* (ACR 7) for Guaymas, Mexico, where their assignment was to convince the Yaqui Indians to cease disturbing American nationals residing in that part of the country. The companies returned to the fair on 10 August 1915 after determining that a landing in Mexico would not be necessary.

The city of San Diego added further incentive to Marine Corps expansion in the fall of 1915, when officials offered the Navy Department 500 acres of municipally owned tidelands on the condition that the military branch purchase an adjoining 232 acres of privately owned lands. By January 1916, Secretary of the Navy Josephus Daniels and the Navy General Board approved a plan to proceed with the real estate purchase and installation of a permanent Marine Corps post at San Diego. Along with Mare Island Naval Shipyard, California, the base became home to the 4th Regiment, and Major William N. McKelvy assumed command of the newly established Marine barracks.

Less than six months later, the 4th Marine Regiment departed San Diego for expeditionary duty in Santo Domingo. On 6 June 1916, Colonel Pendleton and his men embarked by rail for New Orleans, Louisiana, where they then boarded the USS *Hancock* (AP 3) and set sail for the Dominican Republic. The regiment became one component of the 2d Provisional Brigade and served there in support of the U.S. government from 1916 to 1924.

After serving as commander of U.S. forces in Santo Domingo until October 1918, Brigadier General Pendleton assumed command of Marine Barracks Parris Island, South Carolina, from which post he was later detached to activate Headquarters, 2d Advanced Base Force, at San Diego. He arrived at the Marine barracks on 1 October 1919 and established his new headquarters that same day. As construction was still ongoing at Marine Corps Base San Diego, the post continued to be located at Balboa Park where the barracks detachment had grown from

Aerial view of Marine Corps Base San Diego, 1921.
Official U.S. Marine Corps photo, courtesy of Command Museum, Marine Corps Recruit Depot San Diego

the platoon-size organization left by the 4th Regiment in 1916 to approximately 10 officers and 300 men during the First World War.

Pendleton remained in charge of Marine Corps activities in San Diego during the next five years. On 1 April 1921, the 1st Battalion of the 7th Regiment was reactivated at San Diego as a component of the 2d Advanced Base Force. The following November, the latter was redesignated the 5th Brigade, and on 1 December, the 1st Battalion, 7th Regiment, became the 1st Separate Battalion, 5th Brigade. These organizational changes necessitated the move from Balboa Park to Dutch Flats. The soggy marshland had required significant upgrades, but in 1921 the installation was fully operational and the Marine Advanced Expeditionary Base, San Diego, was established. Less than two years later, the Marine Corps recruit depot for the western United States also relocated to the new base at San Diego. With the consolidation of the Marine Barracks at Mare Island Naval Shipyard with the recruit depot in Southern California, San Diego became the sole training site for all recruits from the western part of the United States.

Receiving the official designation of Marine Corps Base, Naval Operating Base, San Diego, on 1 March 1924, the installation largely reflected the vision and leadership of Major General Pendleton. After 46 years of military service, the base commander stepped away from an active role in Marine Corps affairs. He retired in June 1924, secure in the knowledge that the Marine Corps Base San Diego had established a prominent and permanent Marine Corps presence on the West Coast. Major General Pendleton remained an active and public-spirited citizen of Coronado, across the bay from San Diego, and witnessed continuing military development in the region until his death nearly 18 years later on 4 February 1942.

WORLD WAR II
1941–45

With the looming specter of war in 1940, the Marine Corps doubled in size. Prior to World War II, Marines trained primarily in Quantico, Virginia, and Parris Island, South Carolina. A contingent also trained at the Marine Corps Base San Diego (now known as Marine Corps Recruit Depot, San Diego). Although these existing facilities expanded to support the influx of new recruits, the anticipated involvement of U.S. armed forces in the South Pacific precipitated a call for a larger West Coast training installation. Marine Corps Base San Diego was landlocked and thus a poor location for amphibious assault training exercises. A 9,000-acre temporary camp, known as Camp Elliot after former Commandant George F. Elliott, was established north of the city, but that facility, too, soon reached capacity. The Marine Corps began pursuing an additional site for the purpose of training recruits for the Pacific theater.

In the years leading up to World War II, both the Marine Corps and Army expressed interest in acquiring all or part of the Rancho Santa Margarita y Las Flores for training troops. The tract, located approximately 50 miles north of San Diego near the community of Oceanside, offered ample coastline and relatively undeveloped terrain. Split among several private owners and operated by managers for the heirs, the ranch encompassed more than 260,000 acres in San Diego and Orange Counties. Army planners initially envisioned the site as a training base for developing an armored division, but decided that the steep canyons and hills were ill-suited for units preparing to fight in North Africa or Central Europe. The Marine Corps, by contrast, appreciated the property's vast acreage and varied terrain. In addition, its beaches were conducive to amphibious exercises, a particularly import-

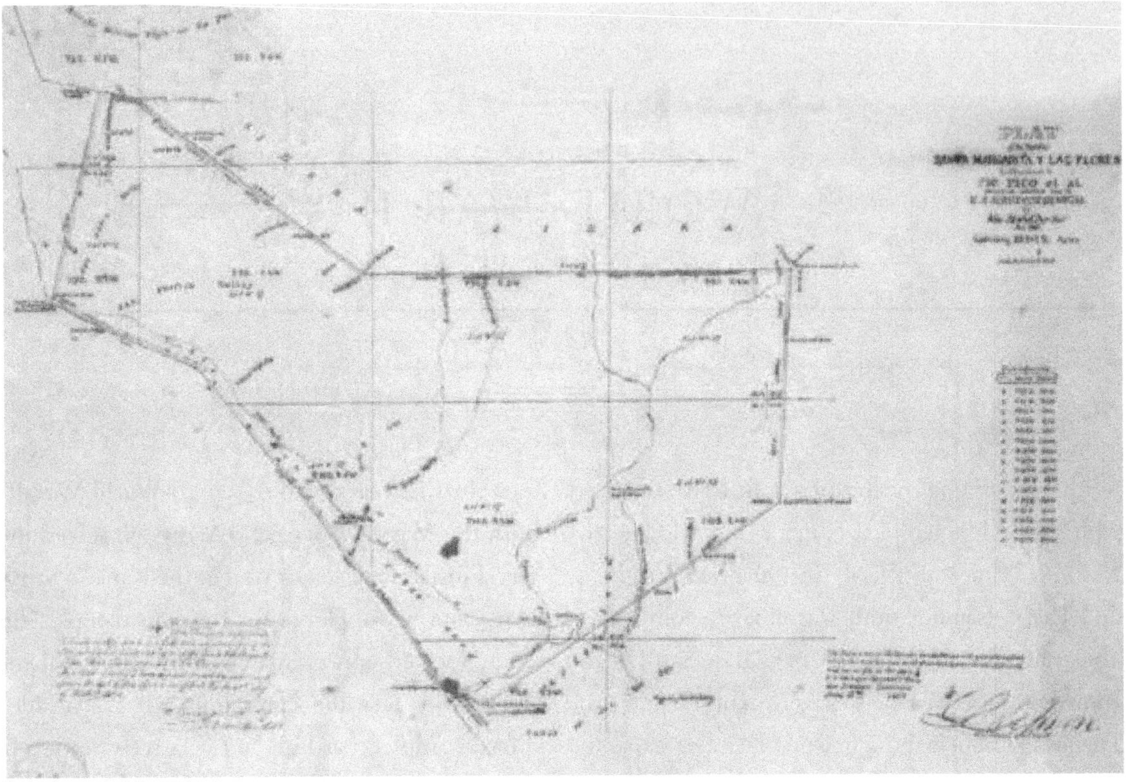

Map of Rancho Santa Margarita y Las Flores by U.S. Deputy Surveyor Max Strobel, July 1867.
The Bancroft Library, University of California Berkeley

ant feature to the Marines due to the greater emphasis being placed on amphibious warfare. Only its distance from San Diego and anticipated development costs prevented Marine Corps planners from recommending the ranch for Marine Corps expansion.

The Japanese attack on Pearl Harbor and U.S. declaration of war in December 1941 added a sense of urgency to military efforts to acquire civilian land for the war effort. Under the Second War Powers Act, the federal government possessed the ability to seize private property for immediate use while it negotiated a fair price with the owners. In April 1942, the military compensated the O'Neill and Flood families a combined total of $4.2 million for approximately 121,000 acres of land, including almost 20 miles of coastal property.[3] This purchase included the core buildings of the Rancho Santa Margarita y Las Flores: the Santa Margarita Ranch House, bunkhouse, and blacksmith shop (now a chapel), as well as the Las Flores Adobe ranch house constructed by Marco A. Forster in 1867.

Military officials initially intended the new training facility to be ready for occupancy in six months and moved quickly to build the basic infrastructure necessary to support 20,000 Marines. More than 5,000 laborers worked around the clock to prepare airfields, barracks, railroads, administration buildings, and other support facilities. The rapid pace of development and large number of expected recruits necessitated the widespread construction of temporary structures. The primary troop facilities included Navy B1-type wood frame barracks with battalion mess halls and tent camps. Storage buildings rested on concrete foundations, while most wood frame administration buildings and barracks of temporary construction were elevated on post-and-pier foundations. Nevertheless, the expedited pace of construction was stymied by changing design plans, the challenge of coordinating multiple survey parties, and spring flooding of the Santa Margarita River and smaller streams running through base. To ensure the military installation would be operational by September, the project managers adopted a vertical approach, focusing construction efforts first on the "Mainside" area in the southeast portion of base and expanding to other designated areas in order of priority.

By late summer, Colonel Lemuel C. Shepherd Jr. led the 9th Marine Regiment with the 1st Battalion, 12th Marines, on a four-day tactical march north from Camp Elliot, east of Miramar, to become the first troops to occupy the new installation. Colonel Shepherd conceived of the multiday trek as an extended training exercise. Having dispatched one of his companies in advance to plan simulated attacks against the marching Marines, the commander ordered his troops to remain on constant alert as though they

[3] The Marine Corps aimed to purchase the land outright, but the Baumgartner family, who had married into the O'Neill family, offered to lease the property to the Marine Corps instead. The Marine Corps deemed it essential to own the ranch so that it could terminate the existing leases with tenants and use the land as it wished. Condemnation proceedings began for the acquisition of 113,000 acres, and the land was vested to the Marine Corps in December 1942. While some families were able to maintain individual leases from the government to continue farming the land until the 1960s, the majority of tenant farmers moved away. In addition, the residents of a small Japanese settlement were relocated to internment camps. Later purchases expanded the total acreage. See Ryan Jordan, "Remembering the Forgotten Village of San Onofre: An Untold Story of Race Relations," *Journal of San Diego History* 60, no. 1–2 (Winter/Spring 2014): 27–44.

Panoramic view of area 1, where the 1st Marine Division Headquarters building is now located.
J. E. Haddock Scrapbook, MCB Camp Pendleton Archives

were traversing enemy territory. The perceived threat that the Japanese might attack Southern California made this pretense easy to follow, and the regiment successfully defeated a series of mock ambushes en route to their new quarters.

President Franklin D. Roosevelt oversaw the official dedication of the base on 25 September 1942. In a simple ceremony honoring the memory of Major General Joseph H. Pendleton, the president designated the new military installation Camp Pendleton after the Marine Corps general's longtime advocacy for a permanent Marine presence in San Diego. Roosevelt was joined by General Pendleton's widow, Mary Fay Pendleton, who affixed the national colors and slowly raised the flag, and Major General Joseph C. Fegan, the first commanding general of Camp Pendleton, who read aloud the orders that formally placed the base in commission.

Designed to provide large-scale tactical training for organizations before they were shipped out to the Pacific, the facility was a geo-

CAMP PENDLETON CONSTRUCTION STATISTICS

Prime contractors:	J. E. Haddock, Ltd. Engineers, Ltd.
Construction duration:	16 months
Cost:	$35 million (initial contract)
Employees:	5,500 (peak) from San Diego and Los Angeles
Construction area:	Pacific Ocean to Riverside County line and between the city limits of Oceanside and San Clemente
Elevations:	Sea level to more than 3,000 feet above sea level
Initial facilities built:	518 buildings consisting of 4.5 million square feet (approximately 100 acres)
Roads built:	14 miles of paved four-lane highway 30 miles of secondary paved roads 65 miles of unpaved roads

The lumber yard at Camp Pendleton, 1942.
J. E. Haddock Scrapbook, MCB Camp Pendleton Archives

graphical model of the concept of a land-sea-air team. Tanks churned in the meadows while infantry trained in the hilly interior wilderness. Amtracs rolled into Del Mar Boat Basin and out to sea, and the new airstrip near the Santa Margarita Ranch House facilitated Marine aviators' training for operations in the Pacific theater. By the end of World War II, the entire 3d, 4th, and

Marines from 9th Marine Regiment and its attachments marched from Camp Elliott to the new Marine Corps base, Camp Pendleton, over four days starting on 1 September 1942.
Official U.S. Marine Corps photo

President Franklin D. Roosevelt and Mary Fay Pendleton, widow of MajGen Joseph H. Pendleton, at the dedication ceremony for Camp Pendleton, 25 September 1942. In the foreground, GySgt Carl A. Ostrom holds the U.S. flag prior to its raising as part of the inauguration of the base.
Official U.S. Marine Corps photo

5th Divisions, or fully one-half of the Marines who fought in the Pacific, passed through the "Gateway to the Pacific." Meanwhile, Camp Elliot had become home to the Fleet Marine Force Training Center, West Coast, and carried the principal mission of training individual replacements for combat units overseas. When the Navy took over the facility in 1944, Headquarters, Fleet Marine Force, San Diego Area, transferred to Camp Pendleton. The merger distinguished the latter as the largest Marine Corps base in the nation. Consonant with its prominence in both population and amphibious training capabilities, Camp Pendleton became a permanent training installation on 14 October 1944.

After the war, Camp Pendleton reversed direction and served as a demobilization center for the Marine Corps. All organized units returning to the United States from the Pacific theater passed through Southern California. Camp Pendleton's Training Command became the Redistribution Regiment, processing 50,000 men

U.S. Marine Raiders dash at full speed along narrow poles as part of training at Camp Pendleton, January 1943.
Official U.S. Marine Corps photo

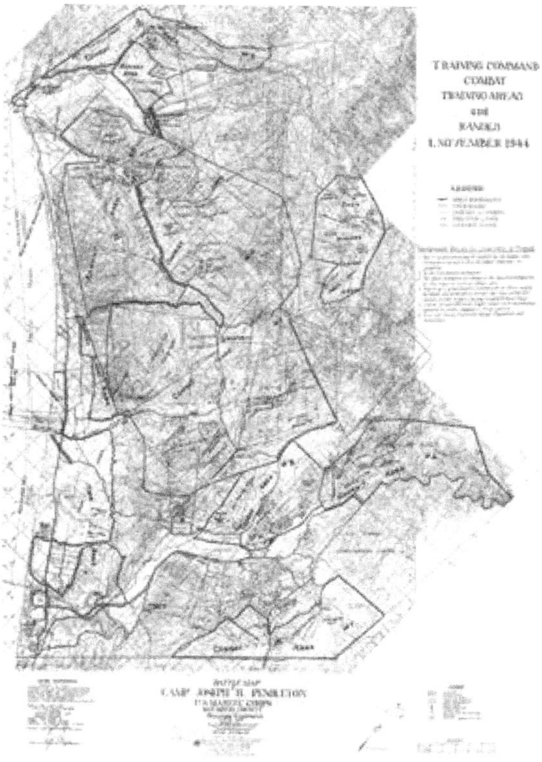

Map of training facilities at Camp Pendleton, 1944.
MCB Camp Pendleton Archives

either for immediate separation or relocation to another station for discharge from the Marine Corps. Several smaller units based at the installation, including the Navajo Code Talkers and the Marine Corps Women Reserve, were disbanded altogether. The total strength of the Marine Corps decreased from 485,000 at the close of the war to 80,000 in 1947.

POST-WORLD WAR II ERA
1946–49

The immediate postwar years at Camp Pendleton witnessed significant organizational changes. In 1946, General Alexander A. Vandegrift, Commandant of the Marine Corps, declared that the base would remain the center of all Marine Corps activities on the West Coast. Furthermore, the military post would be permanently maintained as the home of the 1st Marine Division and Signal School. As the battalion moved from Camp Lejeune, North Carolina, to reduce the concentration of training on the East Coast, the name of the West Coast facility changed from Marine Training and Replacement Command, San Diego Area, to Marine Barracks, Camp Pendleton. With this shift in designation, Camp Pendleton became an independent command, distinct from the San Diego area commands, when Major General Graves B. Erskine assumed leadership of base the following year.

In 1947, President Harry S. Truman signed the National Security Act, which established the Marine Corps as a permanent part of the military attached to the Navy. In so doing, the act formalized the special amphibious function of the Marine Corps. Specifically, its mission under the new organizational structure became defined as the seizure and defense of advanced bases and engagement in land operations related to naval campaigns.[4] These developments secured the future of the Marine Corps and clarified its role within the U.S. armed forces.

[4] The National Security Act also established a separate Air Force and organized the military under a single secretary of defense.

Commandant of the Marine Corps Alexander A. Vandegrift declared that Camp Pendleton would be the center of all West Coast Marine activities and the home of the 1st Marine Division.
Official U.S. Marine Corps photo

The National Security Act of 1947 merged the Department of War and the Department of the Navy into the National Military Establishment. It also created the Department of the Air Force and protected the Marine Corps as an independent Service. In 1949, President Harry S. Truman signed an amendment to the legislation, which converted the National Military Establishment into a new Department of Defense.
Harry S. Truman Library, National Archives and Records Administration, courtesy of Abbie Rowe

KOREAN WAR
1950–53

Regional wars took on new significance during the Cold War as the United States and its allies engaged in military action in distant locales, such as Korea and Vietnam. On 25 June 1950, North Korean ground and air forces cut across the 38th Parallel in a surprise invasion of South Korea. The United States and a coalition of nations responded to a request from the United Nations to enter the conflict in defense of South Korea and to prevent the spread of Communism to new locales.

The Korean War prompted an acceleration of recruitment and training, such that approximately 200,000 Marines passed through Camp Pendleton en route to Asia. The 1st Provisional Brigade with troops from the 5th Marines were the first unit to be trained and shipped out of the installation, all within days of Army General Douglas MacArthur's initial call for a Marine Regimental Combat Team and a supporting air group in July 1950. The 5th Marines arrived in the Pusan Perimeter on 2 August, and the rest of the division followed soon after. The Training and Replacement Command managed the influx of reservists reporting to Camp Pendleton for active duty, while the 1st Advanced Infantry Training Regiment provided four weeks of combat training for boot camp graduates. Special training areas designed to simulate the conditions of fighting in Korea were developed as well, including those on Camp Pendleton, the nearby Marine Corps Mountain Warfare Training Center, and Twentynine Palms.

For the Marine Corps, the Korean War established a pattern of Cold War combat action in regional, limited, and costly wars. The conflict marked the first occasion in which the Marine Corps employed both helicopters and jet aircraft in combat. In addition, the demonstrable suc-

Members of a .30-caliber machine gun platoon of the 1st Provisional Marine Brigade prepare to embark for the Far East, 12 July 1950.
Official U.S. Marine Corps photo, Historical Reference Branch, Marine Corps History Division, Quantico, VA, courtesy of Sgt F. J. Hotman

cess of Marine brigade reinforcement of Army troops at the Pusan Peninsula and the 1st Marine Division's amphibious landing at Inchon and recapture of Seoul helped secure a future for the Marine Corps against critics who argued against a permanent and independent role for the Service. In June 1952, Congress demonstrated its confidence in the Marine Corps by passing an amendment to the National Security Act. The legislation delineated the place of the Marine Corps as a separate Service with distinct roles and missions and guaranteed a force strength of at least three combat divisions and three aircraft wings. General Lemuel Shepherd, who became Commandant on 1 January 1952, set about implementing the law immediately. Camp Pendleton became the headquarters for the 1st and 3d Marine Divisions, which were reformed and

A U.S. Marine patrol closes in on a Korean hut in search of an enemy sniper.
Official U.S. Marine Corps photo, Historical Reference Branch, Marine Corps History Division, Quantico, VA, courtesy of TSgt Frank W. Sewell

reactivated in response to the ongoing war. The 1st Marine Division was swiftly assembled and sent to Korea, while the 3d Marine Division deployed as a replacement force to Japan. The conflict ended in 1953 with a cease-fire that reestablished the previous division of the peninsula along the 38th Parallel.

Following the Korean War, military strategists advocated the development of a more flexible and mobile Marine Corps that comprised divisions capable of being airlifted for combat. The 1st Marine Division remained at Camp Pendleton as the Corps' force in readiness, as the undeveloped coastline and inland areas permitted military leaders ample opportunity to test and refine these reorganization ideas through large-scale exercises in the 1950s and 1960s.

Amphibious exercises became commonplace at Camp Pendleton during the 1950s and 1960s.
Historical Photograph Collection, MCB Camp Pendleton Archives

STAFF SERGEANT RECKLESS
(ca. 1948–68)

Reckless, a small chestnut-colored horse with a white blaze on her face and three white stocking feet, served bravely with the Marine Corps during the Korean War. Recruited into service in October 1952, the mare endeared herself to the Marines through her heroic actions on the battlefield. The Marine Corps honored her service with official rank in the United States military, and in 1959, Reckless attained the rank of staff sergeant under the auspices of General Randolph McC. Pate, Commandant of the Marine Corps.

The distinguished career of this warhorse began when Lieutenant Eric Pedersen, the commanding officer of the Recoilless Rifle Platoon, Antitank Company, 5th Marine Regiment, recognized the value of having a horse to help carry ammunition for his platoon's recoilless rifles. Introduced in World War II, the weapon could be carried by three or four men and project a 65mm shell several thousand yards with precision; however, rounds weighed approximately 24 pounds, which meant that a Marine ammunition carrier was limited in how many he could transport. With two other Marines, Pedersen visited the Seoul racetrack and paid $250 of his own money to buy a little red horse from a young Korean, Kim Huk Moon, who wanted to buy an artificial leg for his wounded sister. When the Marines arrived back at camp with the horse riding in a jeep trailer, members of the platoon gathered to meet the new equine recruit. They dubbed her "Reckless," after the nickname Marines had given the recoilless rifle.

PFC Winford G. McCraken, Headquarters Company, 5th Marines, stands by as Gen Randolph McC. Pate, Commandant of the Marine Corps, tacks chevrons on the blanket of SSgt Reckless, August 1959.

Official U.S. Marine Corps photo, Historical Reference Branch, Marine Corps History Division, Quantico, VA, courtesy of Rhoades

Pedersen assigned Platoon Sergeant Joseph Latham the responsibility of overseeing battlefield training for the horse. Latham taught Reckless how to get in and out of a jeep trailer, how to take cover while on the front lines, and how to return to the bunker when incoming rounds hit behind the lines. The platoon sergeant also secured a pack saddle, which permitted the horse to carry loads of 6–10 rounds to Marines on the front lines. Although initially nervous around the live rounds, Reckless settled into the routine of delivering ammunition to the recoilless gun crew. When not on the front lines, she assisted the platoon with stringing communication wire and with carrying grenades, small-arms ammunition, rations, sleeping bags, and even barbed wire.

Reckless served in numerous combat actions during the Korean War. Her primary duty was to carry rounds for the recoilless rifle platoon from the ammunition supply point to their various firing sites. In March 1953, Reckless demonstrated both courage and grit during the battle at Outpost Vegas. Crossing rice paddies and up a steep 45-degree mountain trail, the little red mare made more than 50 trips from the ammunition supply point to the firing sites.

Despite being wounded twice during battle, Reckless carried more than 9,000 pounds of ammunition, walked more than 35 miles through exposed terrain, and returned carrying wounded soldiers to safety. For her accomplishments during the action at Outpost Vegas, General Pate, commander of the 1st Marine Division, awarded Reckless a battlefield promotion from corporal to sergeant in a formal ceremony, complete with reviewing stand, on 10 April 1954, several months after the war ended. The horse was promoted again to staff sergeant on 31 August 1959 at Camp Pendleton. Pate, then Commandant of the Marine Corps, presided over the ceremony. Reckless was honored with a 19-gun salute and a parade of 1,700 Marines from her wartime unit.[5]

In addition to attaining the rank of staff sergeant, Reckless received numerous military decorations, including two Purple Hearts, a Marine Corps Good Conduct Medal, a Presidential Unit Citation with bronze star, a National Defense Service Medal, a Korean Service Medal, a United Nations Korea Medal, a Navy Unit Commendation, and Republic of Korea Presidential Unit Citation, all of which she wore on her red and gold blanket. In 1971, the 1st Marine Division Association honored Reckless by erecting a memorial at the entrance to the base stables. More recently, the Marine Corps has installed memorial statues by sculptor Jocelyn Russell at Camp Pendleton as well as at the National Museum of the Marine Corps in Quantico, Virginia.

[5] The number of guns in a salute signifies the importance of the event or person(s) being recognized. Nineteen guns are typically reserved for the vice president and American or foreign ambassadors. The only salute higher, 21 guns, is used to honor the U.S. flag, the American president, presidents from other nations, or visiting members of a royal family.

VIETNAM WAR
1963–75

When civil war broke out between the U.S.-backed South Vietnamese government and the Communist regime of North Vietnam and its supporters, known as the Viet Cong, the United States authorized ground troops to support South Vietnamese forces. With its location on the Pacific Coast, Camp Pendleton once again became a springboard for Marines bound for duty in the Far East. Each month, approximately 6,000–8,000 Marines trained at Camp Pendleton for duty in Vietnam.

Marine Corps expansion during the Vietnam War spurred additional growth at Camp Pendleton, the primary training facility and staging installation for operations in Asia. In 1964, the military base established new weapons ranges to alleviate crowding at the Marine Corps Recruit Depot, San Diego, where civilian development was encroaching on the existing training ranges at La Jolla, California. U.S. participation in the Vietnam conflict also generated a severe manpower shortage that necessitated the reactivation of the 5th Marine Division at Camp Pendleton in 1967–68. As the 1st Marine Division had already departed California for Vietnam in March of 1966, the 5th Marine Division was ordered to serve as the permanent West Coast combat-ready force. The division placed units in Hawaii, Okinawa, and California as part of its reinforcement and strategic reserve missions supporting Marines actively fighting in Vietnam.

At peak U.S. involvement in 1969 and 1970, Camp Pendleton ushered close to 80,000 Marines through deployment training each year. After boot camp, Marines would travel to Camp Pendleton where they would be assigned to companies. Each company would spend two weeks studying individual and advanced combat train-

Vietnam village, Las Pulgas, Camp Pendleton.
Historical Photograph Collection, MCB Camp Pendleton Archives

Marines participated in counterinsurgency training at the simulated Vietnam village aboard Camp Pendleton.
Historical Photograph Collection, MCB Camp Pendleton Archives

ing, basic infantry training, and combat familiarization. Those receiving specialized training, such as communications, field medical, and language studies, would then advance to technical training in the Schools Battalion. Marines assigned to infantry units completed an additional two weeks of training as riflemen, machine gunners, mortar men, scout snipers, and reconnaissance personnel. To prepare new recruits for the particular challenges of guerrilla warfare in Vietnam, the Korean village combat town was converted into a Vietnamese village, complete with booby traps. After training, Marines would spend 15 days at the Staging Battalion, where they received dog tags, clothing, ID cards, immunizations, and physical and dental examinations before departing for Vietnam or stations overseas. Additional training involved locating and disarming booby traps, identifying Communist Vietnamese troops, and firing and maintaining the M16 rifle.

President Richard M. Nixon began a gradual withdrawal of U.S. forces from Vietnam in 1969. Six years later, after the United States officially pulled out and North Vietnamese forces began to encircle Saigon, the U.S. government initiated one of the largest humanitarian airlifts in history. In a matter of days, American helicopters evacuated thousands of U.S. military personnel and their families, as well as Vietnamese allies and other foreign nationals whose support of the United States placed them in grave danger. The displacement of Vietnamese refugees precipitated Operation New Arrivals, a resettlement program consisting of processing centers and tent camps at four military bases on the U.S. mainland, including Camp Pendleton. Housed in large squad tents in a relocation center at Camps Cristianitos and Talega, more than 36,000 refugees passed through the base during the first two phases of the operation. An average of 210 refugees arrived each day, and an average of 262 departed daily for resettlement. The final phase began in September with the last refugee departing Camp Pendleton on 31 October. Over the course of the operation, Marines erected more than 1,000 tents and processed more than 50,000 refugees. The camp officially closed on 15 November 1975.

Refugee children continued their studies at Camp Pendleton during Operation New Arrivals. Many attended their first classes in English at this time.
Official U.S. Marine Corps photo, courtesy of Gill

POST-VIETNAM ERA 1976–PRESENT

After Vietnam, the Marine Corps shifted its mission capabilities from amphibious tactics to expeditionary warfare, with a particular focus on special operations and urban warfare. In the 1980s, Camp Pendleton became the West Coast home of the U.S. Navy's new landing craft air cushion (hovercraft) and Assault Craft Unit 5, the purpose of which was to support naval operations by transporting troops, food, and firepower via coastal landings. Combining infantry, armor, supply, and air power, Marines have demonstrated the effectiveness of the air-ground team to meet modern exigencies in both humanitarian and combat operations.

In the summer of 1990, the 1st Marine Division deployed to the Persian Gulf in Operations Desert Shield and Desert Storm. Comprising part of the largest deployment of U.S. military forces since Vietnam, the Marine Corps underscored the speed and efficiency of American military might through its rapid deployment of troops and equipment from Camp Pendleton to the Middle East in a matter of days. The division returned to the United States by April 1991.

Today, Camp Pendleton is home to the I Marine Expeditionary Force (I MEF) and its subordinate units, as well as tenant units at Marine Corps Air Station Miramar, Marine Corps Tactical Systems Support Activity, and the Navy's Assault Craft Units. The base further provides specialized schools and training as directed by the Commandant of the Marine Corps. The military base remains an outstanding training facility preparing military personnel for rapid deployment and maintaining combat readiness in the twenty-first century. Its vast acreage allows the use of modern weaponry and permits trainees to engage in landing exercises and tactical maneu-

The 1st Marine Division formed the first combat-ready forces in Saudi Arabia. In early August 1990, Marines at Camp Pendleton learned that Iraqi forces led by Saddam Hussein had invaded Kuwait and immediately began preparations for deployment.
Official U.S. Marine Corps photo, courtesy of GySgt Chuck Archuleta

vers. With its varied terrain, Camp Pendleton also supports experimentation for a wide range of combat operations. Since the 1980s, Marines stationed at Camp Pendleton have served in Grenada, Panama, the Persian Gulf, Somalia, and the wars in Iraq and Afghanistan.

Because Camp Pendleton is the home of I MEF, 1st Marine Division remains at the forefront in the Global War on Terrorism, maintaining 14 military operations in urban terrain facilities to prepare Marines for combat in urban environments.
Official U.S. Marine Corps photo

Reconnaissance Training Company Marines receive an aerial view of Camp Pendleton during Special Patrol Insertion/Extraction training at the San Mateo Landing Zone, 17 April 2015.
Official U.S. Marine Corps photo, courtesy of LCpl Asia J. Sorenson

ARCHITECTURAL FEATURES OF 1ST MARINE DIVISION HEADQUARTERS BUILDING 1133

The 1st Marine Division Headquarters, also known as the "White House," is a World War II-era administrative building. Designed in September 1942 by the architectural firm Hunt, Chambers, and Ellingwood, the two-story building is typical of the utilitarian and swiftly constructed temporary structures that sprang up in response to wartime mobilization during World War II. The rambling wood-frame building takes the shape of two Swiss crosses connected via a long, central façade. A prominent porch spans the principal elevation, connecting north and south wings, and the rhythmic placement and proportioning of window openings (2:1 ratio) create a sense of geometric harmony on an otherwise austere façade. The majority of World War II-era buildings at Camp Pendleton were of a temporary nature, and the 1st Marine Division Headquarters remains one of the few relatively unaltered structures of this type on base. Located in the Mainside area, it has served as the primary administrative building at Camp Pendleton since its construction in 1943. Its surviving architectural features include a sprawling shape, raised pier-and-beam foundation, and wide nine-bay porch.

The White House was an early and important structure in the development of the new West Coast training facility. Its original purpose was to serve as the headquarters building for all Marine Corps divisions assigned to Camp Pendleton. By the end of World War II, the 3d, 4th, and 5th Divisions, or fully one-half of the Marines who fought in the Pacific, used this building as their command post. In 1946, General Alexander Vandegrift, Commandant of the Marine Corps, ordered the base to remain the center of all Marine Corps activities on the West Coast and designated the site as the permanent home of the 1st Marine Division. The White House

1st Marine Division Headquarters, Building 1133, August 2016.
Photo by the author

has been used continuously as the official West Coast headquarters for that division ever since.

Two recent events had a profound influence on the design and setting of the Marine Division Command Post building: the Japanese attack on Pearl Harbor in December 1941 and springtime flooding near the Santa Margarita River in 1942. Apprehension over the possibility of Japanese raids led architects Myron B. Hunt, Harold C. Chambers, and E. L. Ellingwood to abandon a prescribed military master plan for Camp Pendleton in favor of a deliberately random arrangement of buildings. Dispersing structures across the landscape and embedding a shield of vermiculite, or lightweight concrete, under the roofing, the architects intended to protect the military installation from enemy raids by making the pattern of bombing problematic and the spread of

Earliest known photo of Building 1133, taken shortly after construction, ca. 1943.
J. E. Haddock Scrapbook, MCB Camp Pendleton Archives

fire more difficult. Consequently, Hunt's firm discretely sited the headquarters building as a stand-alone structure, removed from the cluster of administration buildings that traditionally marked the center of a base. Today, the 1st Marine Division Headquarters stands relatively isolated compared to newer construction at Camp Pendleton. Set back from one of the primary roads on base, the building overlooks a vast parade ground and remains largely surrounded by open space.

To counter the threat of seasonal weather, the architectural firm raised the administrative building on a pier-and-beam foundation. This adaptation lifted the principal office space above the threat of flooding due to heavy rains. In an era when central air conditioning was not yet commonplace, the design offered the additional benefit of allowing the underside of the building to breathe.

Adapting the design from standardized U.S. Navy B1-type barracks, Hunt, Chambers, and Ellingwood devised plans that upgraded architectural details and incorporated new materials to meet the particularities of function and climate for the Marine Division Command Post. Barracks were generally two-story frame construction with the interior space partitioned into offices, classrooms, and dormitories. With a double Swiss cross floorplan, the White House is more complex in its layout than standard barracks; yet like those structures, the interior

Building 1133, ca. mid-to-late 1950s.
Collection of LtCol H. J. L. Reid, USMC (Ret), MCB Camp Pendleton Archives

arrangement reflects the external logic of the structure and creates circulation within. Double-wide entrances at each end of the porch open to a small entry hall with a wide stairway to the second floor. These modest lobbies lead to a central corridor system connecting all segments of the building. On the first and second floors, as well as in each side wing, openness is achieved through long, narrow hallways flanked by a single row of rectangular rooms on either side. The second floor contains a few larger rooms, reserved for the most important occupants of the building. Otherwise, there is little to distinguish any hierarchy of space throughout the structure.

Embodying characteristics of Hunt, Chambers, and Ellingwood's temporary World War II buildings, the White House displays few overt stylistic features. It exhibits minimal surface ornamentation and lacks skirting around the foundation crawlspace. The vernacular aesthetic of the structure is further expressed in the architects' near-symmetrical massing of form and in their reliance on simple building materials such as wood. The open railing, low gabled roof, and wood sheathing further associate the building with temporary structures erected during the war mobilization years of the early 1940s. These defining characteristics remained unchanged un-

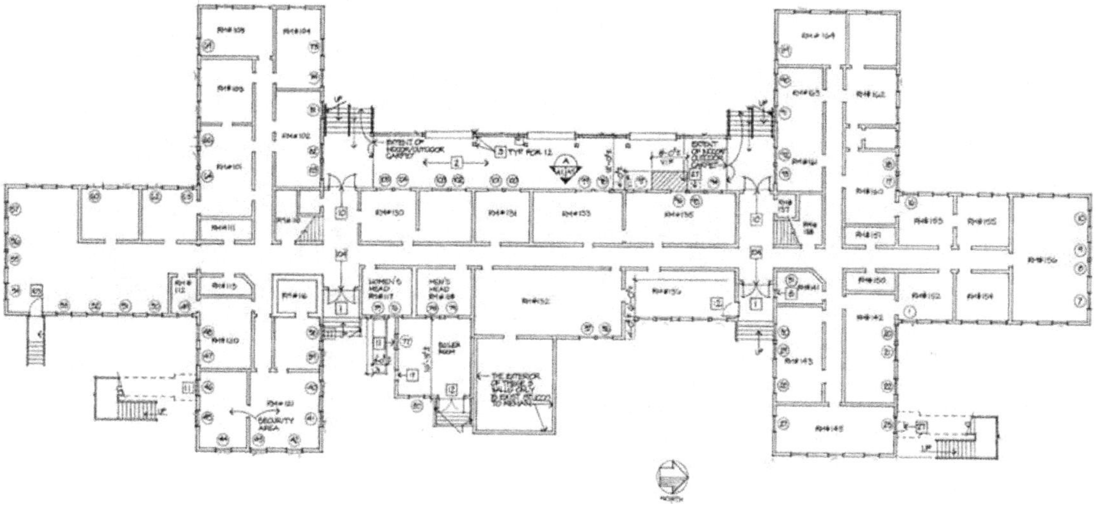

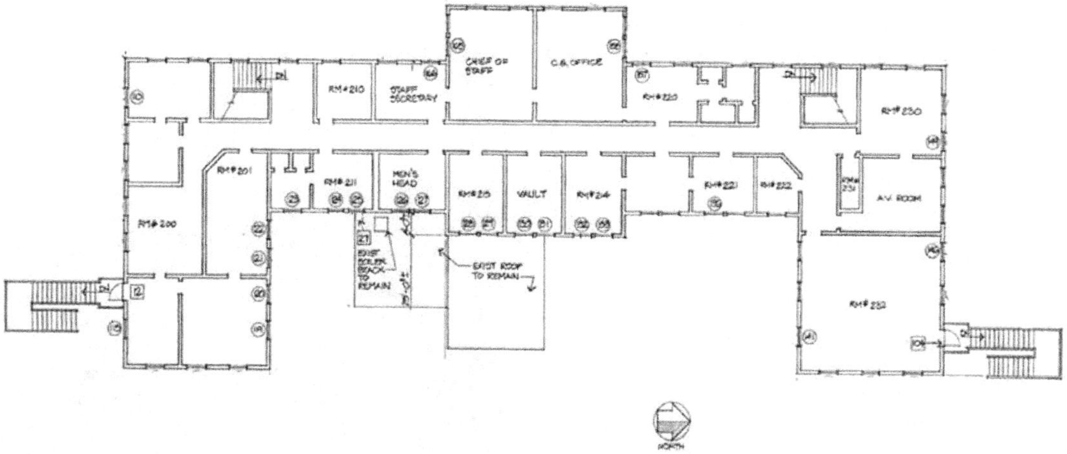

Top: first-floor plan of Building 1133. Bottom: second-floor plan of Building 1133.
ASM Affiliates, Carlsbad, CA

This photograph of the 1st Marine Division Headquarters, taken in the 1970s or 1980s, shows the modern screen alterations made to the porch during the 1960s.
Historical Photograph Collection, MCB Camp Pendleton Archives

til the 1960s, when a major renovation altered the building's exterior appearance. At that time, the primary entrance, which the architects had centered along the symmetrical front façade and accentuated with a "sign marker" or prominent center bay, was enclosed and the center bank of steps eliminated from the front porch. The porch railing system also changed significantly as part of this construction; the horizontal rails between posts were replaced with a vertical screening device. Although typical of mid-century modern architecture, the visual obstruction of the porch and diversion of foot traffic to secondary side entrances nevertheless diminished the structure's original legibility of function and transparency of design.

A second renovation project in 1983 aimed

New millwork replaced the original in the interior hallways and stairwells of the 1st Marine Division Headquarters.
ASM Affiliates, Carlsbad, CA

The porch railing of the 1st Marine Division Headquarters has been restored to handrail height.
Photo by the author

to refresh the interior of the White House. This undertaking involved laying carpet over the original flooring and painting the walls a deep forest green color. In addition, many doors were replaced with colonial-style raised panel units, while ornamental oak millwork embellished door trim, stairway facings, and walls to create the impression of a wainscot. These decorative elements—the Victorian-style millwork, colonial-style doors, paint colors, and carpeting—represent a significant departure from the building's minimalist wartime inception. Moreover, many areas feature suspended ceilings,

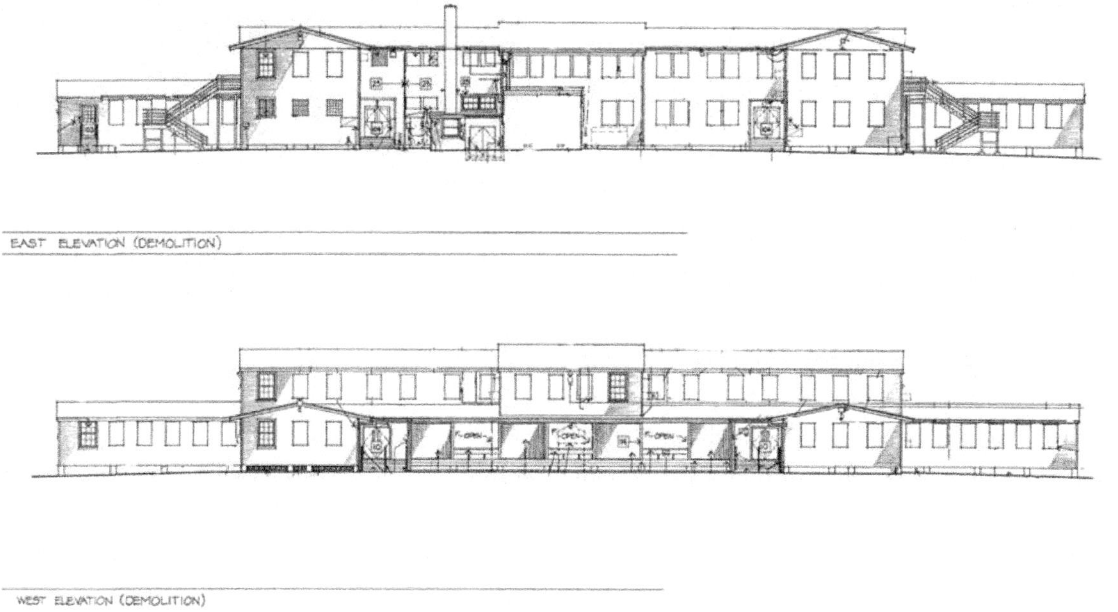

Main elevation (West and East) of Building 1133.
ASM Affiliates, Carlsbad, CA

which cover the original acoustic tile-clad ceilings and reduce the overall height of corridors and rooms.

Subsequent additions and updates involved the replacement of wood windows with aluminum units, removal of original window sills and trim, overlay of the original German lap siding with fiberboard product, and enclosure of the foundation-level crawl space with framed skirts. A partial remodel of the front porch railing system restored the height of the vertical screening to handrail height, such that the veranda currently contains alternating bays characterized by vertical louvers and planter boxes. In addition, the low-profile gable roof was updated with asphalt shingles, although the gentle pitch and material remain consistent with the original design of the structure.

Today, the White House remains a tangible example of the growing presence of the Marine Corps on the West Coast during World War II. Despite extensive surface modifications, the floorplan and configuration of space at the 1st Marine Division Headquarters has changed little since its construction in 1943. The uniqueness of its sprawling shape, color, wide porch, and fenestration all combine to evoke a strong sense of time and place. In 2011, a historic resources survey and evaluation at Marine Corps Base Camp Pendleton recommended the Marine Division Command Post building for eligibility with the National Register of Historic Places for its significance as a representative example of temporary military base architecture from the World War II-era.

Front entrance of the Marine Division Command Post showing the blue diamond insignia of the 1st Marine Division. The first shoulder patch to be approved during World War II, the blue diamond insignia commemorates the division's sacrifices and victory in the Battle for Guadalcanal.
Photo by the author

MYRON H. HUNT

Myron H. Hunt (1868–1952) was one of the most prominent Southern California architects of the early twentieth century. With a career spanning five decades, he designed numerous well-known California landmarks and upper-class homes using an eclectic range of styles.

Born in Sunderland, Massachusetts, Hunt spent his formative years in the Lakeview district of Chicago, Illinois. He graduated from Lake View High School and briefly attended Northwestern University before advancing to the Massachusetts Institute of Technology, where he earned a degree in architecture in 1893. He then extended his education with a three-year sojourn to Europe, where he studied architectural antiquities. Upon his return to the United States, Hunt joined the Chicago office of Shepley, Rutan, and Coolidge, a prominent Boston-based architectural firm. During this time, Hunt was chiefly involved in the design of expensive suburban homes, for which he called upon Chicago architect Frank Lloyd Wright's innovative Prairie School aesthetic for inspiration.

Architect Myron H. Hunt (1868–1952). This portrait was taken by Master Studio, NY, in 1927.
Palos Verdes Library Local History Room Photograph Collection, Palos Verdes Library District

In 1903, Hunt left behind a highly successful practice in Chicago and moved to Pasadena, California, to ease his wife's suffering from tuberculosis. During the next 20 years, he maintained an office in Los Angeles, where he worked both independently and in partnership with fellow Midwestern architect Elmer Grey (1872–1963). Among Hunt's best-known commissions of this period are Henry Huntington's estate (now the Huntington Library, Art Collections, and Botanical Gardens) in San Marino; the Rose Bowl Stadium in Pasadena; the Ambassador Hotel in Los Angeles; and the campus of Occidental College in Los Angeles.

Hunt formed a new partnership with architect Harold C. Chambers (1885–1971) in 1921. A graduate of the Armour Institute of Technology in Chicago, Chambers had worked as a junior architect for the firm Hunt and Grey prior to its

Designed one year before Myron Hunt's retirement, the Beach Club at MCB Camp Pendleton demonstrates a return to the Californian style he had helped to foster. He intended the recreational facility to be sympathetic with the architectural styles of Las Flores Adobe National Historic Landmark and Santa Margarita Ranch House National Historic Site.
MCB Camp Pendleton Archives

dissolution in 1912 and had maintained a close professional relationship with Hunt. The pair operated a successful firm, Hunt and Chambers, until Hunt's retirement in 1947. Their commissions included public libraries in Pasadena, Redlands, Palos Verdes Estates, and Santa Barbara.

After World War I, Hunt's designs evolved into a bold, personal style that he called "Californian," an interpretative take on the Spanish Revival style fusing historical and modern influences. The Beach Club at Marine Corps Base Camp Pendleton, completed in 1946, exemplifies the low gable roof, long veranda, and irregular window and door openings typical of his designs. An avid gardener, Hunt felt that the style grew naturally out of the landscape and culture of Southern California and thus was more appropriate to the region than European precedents. He also intended the Beach

Club, in particular, to harmonize visually with the historic architectural styles of the Santa Margarita Ranch House complex and Las Flores Adobe located nearby.

The 1st Marine Division Headquarters building at Camp Pendleton is a strong example of Hunt's late work, which increasingly focused on public sector projects during the lean years of the Great Depression and World War II. Base planners frequently involved civilian architects in military design projects, and Hunt completed numerous commissions for the U.S. Navy and Marine Corps over the course of his career. With Chambers, he developed site plans, buildings, roads, water supply, sewage, and utilities for several military installations. The pair hired E. L. Ellingwood to assist with temporary construction projects during the war mobilization years. In addition to serving as the original architects for Camp Pendleton, Hunt and his partners received commissions for Naval Air Station North Island and Camp Callan in San Diego; March Field in Riverside; and Camp White in Medford, Oregon. Designed largely as temporary structures, many of the firm's designs have been destroyed or substantially altered since their construction. At Camp Pendleton, the 1st Marine Division Headquarters building remains as the only known example of Hunt's military base design work to be completed during the war.

ARCHITECTURAL FEATURES OF THE SANTA MARGARITA RANCH HOUSE NATIONAL HISTORIC SITE

Among the oldest structures on the grounds of present-day Camp Pendleton are a cluster of adobe buildings belonging to the historic Rancho Santa Margarita y Las Flores, an expansive tract of ranch land whose origins date to the first quarter of the nineteenth century. Located at the corner of Vandegrift Boulevard and Basilone Road, the property includes the main residence and outbuildings now known as the Ranch House, the Bunkhouse Museum, and the Ranch House Chapel. The Santa Margarita Ranch House complex, often referred to simply as the "Ranch House," features adobe-and-timber construction common to Spanish colonial architecture in early California. Placed on the National Register of Historic Places in 1971, the site exudes historical charm with mission-style architectural features including wood-framed ceilings, tiled roofs, covered porches, and plastered walls constructed from locally sourced sun-dried brick.[1]

Constructed between approximately 1810 and 1827, the Santa Margarita ranch house building initially formed one of six *ranchos* providing the Spanish clergy at Mission San Luis Rey with tallow and hides, which could be traded for European goods, such as printed pottery. In an 1827 inventory of the *estancia*, or outpost, Father Antonio Peyri observed a modest two-room structure measuring 44 feet by 36 feet with one

[1] The National Register of Historic Places is the federal government's official list of cultural resources that have been identified as being worthy of preservation. Authorized under the National Historic Preservation Act of 1966, it is part of a nation-wide program to coordinate public and private efforts to identify, evaluate, and protect historic sites in the United States.

door, a tile roof, and a walled enclosure protecting a garden and a vineyard:

> To the North at a distance of one league and a half [about four miles] the mission was a place with a house and a garden, and near the beginnings of the sierra, a vineyard. The land is cultivated and wheat, corn, beans, and barley are raised. The fields are irrigated by means of water from the sierra, which, though not plentiful, assures some crops.[2]

Today, the Ranch House grounds contribute significantly to its historical appeal. Built atop a rocky knoll, the adobe buildings possess a commanding view of the picturesque Santa Margarita River valley.

After Mexico gained independence from Spain in 1821 and Alta California became a possession of the Mexican government, the Santa Margarita Ranch House complex flourished as one of the largest and most successful cattle

Aerial view of the Santa Margarita Ranch House.
Camp Pendleton Historical Society

[2] Stephen D. Mikesell and Stephen R. Wee, *National Register Nomination: Santa Margarita Ranch House* (Washington, DC: National Parks Service, 1994), 3, 25.

Established in 1798, nearly 30 years after Spanish explorer Gaspar de Portolá and his expeditionary force arrived in Alta California, Mission San Luis Rey facilitated Spanish claim-staking in the region through the colonization of indigenous populations. Franciscan priests preached to the Luiseño people, whose villages dotted nearby valleys, streams, and coasts, and taught them agricultural skills to benefit the Spanish Crown. This oil painting, created by artist Christian Jörgensen (1860–1935) depicts Mission San Luis Rey as it appeared in 1910.
The Bancroft Library, University of California Berkeley

ranches in the region.[3] In an effort to disperse the enormous wealth and power that had long been the exclusive purview of the Franciscan clergy, the new Mexican leadership opened the frontier with more liberal trade practices and commenced secularizing the California mission system, including Mission San Luis Rey. Among the beneficiaries of this redistribution process were Andrés and Pio Pico of San Diego. Members of the *diputación,* or local legislative body in Alta California, the Pico brothers were foremost proponents of the liberal movement and

[3] While a Spanish colony, the province of Alta California encompassed the present-day states of California, Nevada, and Utah, as well as parts of four additional states. The Mexican Constitution of 1824 designated this area as Alta California Territory, with Monterey as the capital. The United States occupied much of the area during and following the Mexican-American War and the state of California was formally organized at a constitutional convention held in Monterey in late 1849.

influential leaders in the community. Following the revolution in 1821, Pio, the elder brother, served as one of two commissioners appointed by Governor José Figueroa during the Mexican secularization process to take possession of Mission San Luis Rey and later served as mission administrator.[4]

Pio Pico moved into the old mission building and began erecting for his family a suitably aristocratic abode, the Santa Margarita Ranch House, between 1842 and 1844. Built at or near the site of the older adobe structure, the new residence consisted of a gable-roofed one-and-a-half story main block with an open interior court. The structure was characterized by thick adobe brick walls and giant hand-cut ceiling timbers supporting a heavy, fired clay tile roof and covered porches or *corredors*. A free-standing kitchen wing was oriented at an approximate right angle to the larger structure.

An ensuing cattle boom greatly increased their fortunes, and the Pico brothers added to their land holdings over the next decade. In 1844, the Picos acquired an adjoining parcel at Las Flores, the site of another historic estancia near the village of San Onofre.[5] The combined land grant became known as the Rancho Santa Margarita y Las Flores.

In 1845, Pio Pico entered office as the last

Pio de Jesus Pico IV (1801–94) was a prominent Californio rancher and politician in the Mexican government from 1828 to 1848, when California became part of the United States. Following the Treaty of Guadalupe Hidalgo, Pico became a private American citizen, businessman, and later served on the Los Angeles City Council.
Whittier Museum, Whittier, CA

[4] Although the law of secularization stated that former mission land would transfer to native populations, the terrain instead came under the control of various secular administrators, who awarded large parcels of land to prominent Californio citizens sympathetic to Mexican governance.

[5] Rancho San Onofre y Santa Margarita was the likely original name of the Picos' land.

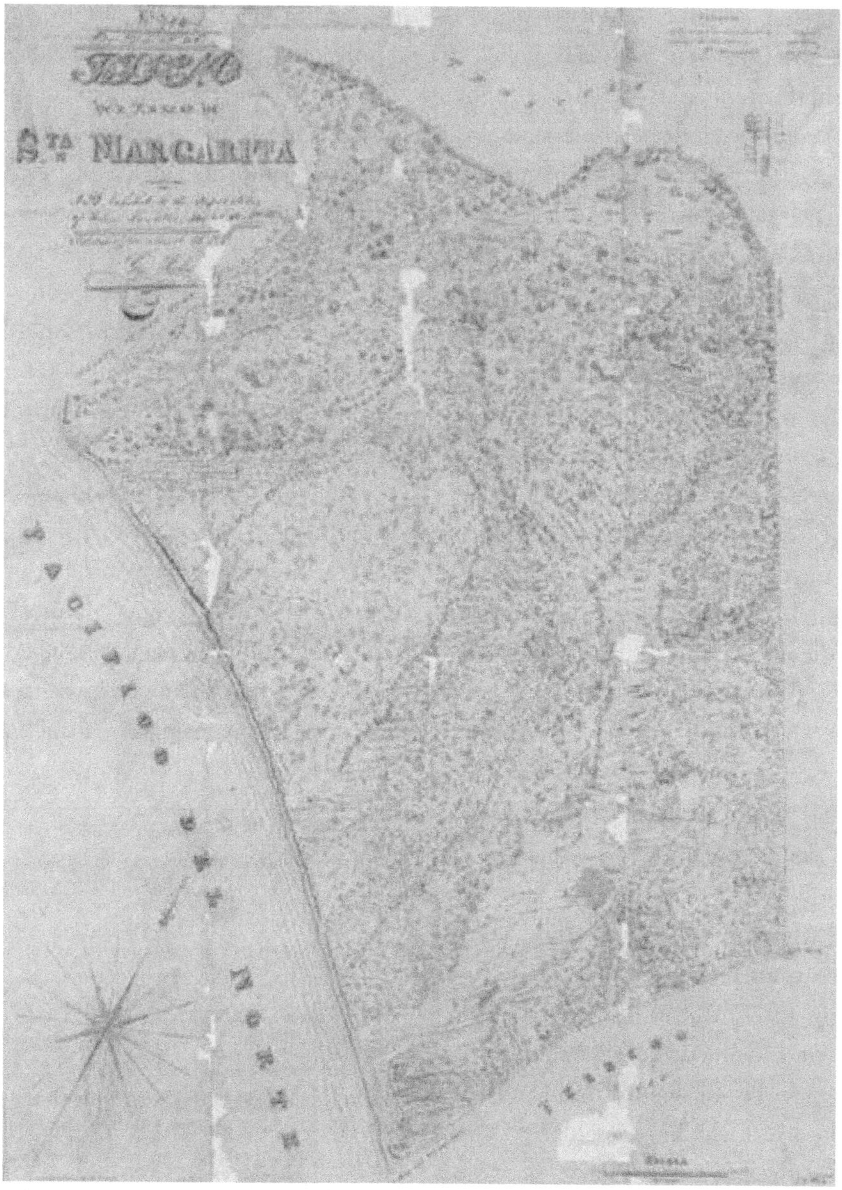

Surveyor and architect Jasper O'Farrell (1817–75) delineated the boundaries of Rancho Santa Margarita in 1845.
The Bancroft Library, University of California Berkeley

Mexican governor of Alta California prior to the Mexican-American War. The arrival of Army Brigadier General Stephen W. Kearny the following year cut short Pico's term in office, and the elder brother fled to Mexico. Remaining in California, Pio's brother Andrés served as a general in the Mexican Army, in which capacity he fought General Kearny at the Battle of San Pasqual in 1846. Remarkably, the ranch remained intact during these tumultuous years, which resulted in the transfer of California sovereignty to the United States. Both brothers returned to the estate after the war, but neither stayed long; Pio relocated to Los Angeles, and Andrés left the region in pursuit of gold in Northern California.

The Rancho Santa Margarita y Las Flores remained in the Pico family until the 1860s, when the brothers trasnferred their shares to their English-born brother-in-law John Forster to avoid foreclosure. Californios were swiftly becoming a minority and facing diminished political power and social status under U.S. governance. Reversals in the livestock trade and the high costs of defending Mexican-issued land grants added further to the challenges facing the Pico brothers. Already in 1862, Andrés Pico found himself deeply in debt and deeded his portion of the ranch to Pio. A severe drought in 1863 and 1864 decimated the livestock, and the elder Pico was forced to sell as well.[6]

Like many of the Anglo merchants who immigrated to Alta California well before the gold rush, Forster embraced Spanish language and customs and married into a prominent Hispanic family. A successful businessman, Forster acquired multiple land grants during Pio Pico's tenure as governor and became a wealthy *ranchero* during the cattle boom. Even before adding Santa Margarita to his holdings, Forster had amassed more than 100,000 acres headquartered at the old Mission San Juan Capistrano and assumed the title "Don."[7]

Rains in late 1864 and early 1865 brought relief from the drought that had forced Pico to sell the ranch, and Forster began making improvements to the property almost immediately. The renovation campaign produced a one-and-a-half story adobe main block with a series of one-story wings enclosing a central courtyard.[8] By

[6] Although Pio Pico offered the mortgage brokers Picche and Boyerque a deed to the property and 5,000 head of cattle as payment, the French entrepreneurs refused. Thus, when his brother-in-law John Forster offered $14,000 in exchange for the deed to the ranch, the elder Pico accepted.

[7] Forster moved from Capistrano to Las Flores in 1864. He attempted to improve his cattle holdings, but his aspirations were stymied by the Picos, who claimed that Forster had purchased only half of the ranch. Andrés purported still to own his original parcel of the estate, and he presented a map as part of his court battle. Forster brought the claim to court, and after an extended trial, the San Diego District Court decided in Forster's favor.

[8] The historical record is unclear on the exact floorplan of the Santa Margarita Ranch House at the time Forster acquired the property. As a result, it is difficult to distinguish which renovations were completed as part of Pio Pico's 1840s construction and which emerged through Forster's subsequent projects.

The Santa Margarita Ranch House, ca. 1880. The vineyard covered 12 acres, while the orchard spanned roughly 5 acres and supported 100 apple and 80 peach trees.
California History Section Picture Catalog, California State Library

The bunkhouse that lies to the south of the Ranch House probably began construction during John Forster's ownership between 1865 and 1870. The building appears in several nineteenth-century photographs, plastered and roofed in tile.
Collection of Jerome Baumgartner Jr., MCB Camp Pendleton Archives

John "Don Juan" Forster (1815–81) rose from modest circumstances in Liverpool, England, to become one of the largest landowners in California. He married Ysidora Pico, sister of Pio and Andrés Pico, in 1837.
San Juan Capistrano Historical Society

the 1870s, the Ranch House included corredors on the west and south sides of the main block, the south side of the south wing, and surrounding the courtyard. As was typical in adobe construction in nineteenth-century California, walls exposed to external elements like sun and rain were originally covered with lime plaster, while interior walls employed a covering of earthen plaster and limewash. In addition, Forster purchased tile from the Mission San Juan Capistrano, his former residence, as roofing material for the new construction. He built a retaining wall on the west side of the building, added a corral and two barns, a winery and a wine cellar, improved the orchards, and established new fields for cultivation.

The Rancho Santa Margarita y Las Flores gained renown as a destination for travelers, who remarked on the Forster family's lavish hospital-

FORSTER CITY

John Forster encountered financial difficulties in the latter part of the nineteenth century, when Southern California saw a shift away from traditional ranching with large cattle herds and the Spanish *padrone* lifestyle to smaller farming operations and fenced properties. By the 1870s, he was looking for a way to profit on the ranch through real estate development. Hoping to emulate the success of Los Angeles ranchers like Abel Stearns, Forster planned to sell off ranch land and create a housing subdivision. By 1880, the town of Forster City could boast of a general store, post office, blacksmith shop, hotel, lumber yard, and livery stable. In time, however, the venture failed.

ity. Recounting a conversation he had had with Forster's son Marco, travel author Charles B. Nordhoff described the reception one could expect to receive at the Ranch House:

> In those days, said Don Marco Forster, men used to travel from San Diego to Monterey and never spend a cent of money. When night came, you stopped at the nearest house. After supper, you were shown your room. In the morning, a clean shirt was at your bedside; and if you were known to the family, it was customary to place near the bed, on the table, also a sum of money, a hundred or two hundred dollars, from which the visitor, if he needed it, was expected to help himself.[9]

The ranch also maintained during this period a relatively large workforce of between 30 and 50 people, many of them servants. Faced with declines in the market, loss of livestock to drought, and high taxes on unimproved land, Forster sought out mortgages with high interest rates to sustain the customary ranchero lifestyle and to prevent foreclosure. Following Forster's death in 1882, his heirs sold the ranch to settle the estate.

In the final decades of the nineteenth century, the Rancho Santa Margarita y Las Flores became one of the largest and most prosperous ranches in California under the enduring partnership of James C. Flood and Richard O'Neill. In addition to the 133,000 acres at Santa Margarita y Las Flores, they purchased two neighboring ranches in Orange County: Mission Viejo and Trabuco. At its peak, the ranch exceeded 205,000 acres, extending from Oceanside to Orange County, and became known simply as the Santa Margarita Ranch.

[9] Charles Nordhoff, *California: For Health, Pleasure, and Residence—A Book for Travelers and Settlers* (New York: Harper & Brothers, 1873), 243.

Oil painting of the Santa Margarita Ranch House, artist unknown.
MCB Camp Pendleton Archives

Whereas Flood had provided the majority of the purchase price for the ranch, O'Neill assumed responsibility over the day-to-day operations. Shortly after the O'Neill family moved into the old Ranch House, the California Southern Railroad completed a line across the ranch. With tracks located immediately north of the Ranch House, long cattle drives to markets in Los Angeles and San Diego became a relic of the past. With new transportation methods came new infrastructure requirements, and O'Neill constructed a small rail station, warehouses, stock pens, and loading platforms in several places throughout the estate. By 1885, he erected a windmill and pump house directly behind the corral. He also ceased the commercial production of wine and converted the winery building into a blacksmith, carpenter, and tool shop.

Born in Mitchelstown, County Cork, Ireland, Richard O'Neill (1824–1910) met James Flood on a ship en route to the United States. The two men formed a lasting friendship and prosperous partnership that included the vast ranching operations at Rancho Santa Margarita y Las Flores. As ranch manager, O'Neill devoted a portion of his yearly profits to Flood to purchase one-half interest in the venture. He became a full partner by 1906.
San Juan Capistrano Historical Society

The adobe portion of the building that is now used as a chapel was constructed by John Forster and likely used as a winery. This photograph, taken during the Flood-O'Neill period, shows the Ranch House building after it had been converted to a blacksmith shop.
Collection of Jerome Baumgartner Jr., MCB Camp Pendleton Archives

By the turn of the century, the Santa Margarita Ranch House had been altered from its 1870s configuration. It is not clear whether the Floods or the O'Neills were responsible for the change in floorplan, or whether Forster had completed some additional remodeling prior to his death. As indicated by an architectural model of the residence from this period, the building contained corredors on the west side of the main block and the south and north sides of the south wing, all covered in clay roofing tiles. The corredors on the remaining three sides of the courtyard and on the south side of the main block were roofed in cedar shingles, suggesting that the corredors denote two distinct construction campaigns. The model also reveals a second-story addition near the southwest corner of the house, which may have served as quarters for the cook, and the enclosure of a portion of the south portal on the kitchen wing.

After O'Neill passed away in 1910, the Flood-O'Neill partnership continued under the guidance of their sons James C. Flood Jr. and

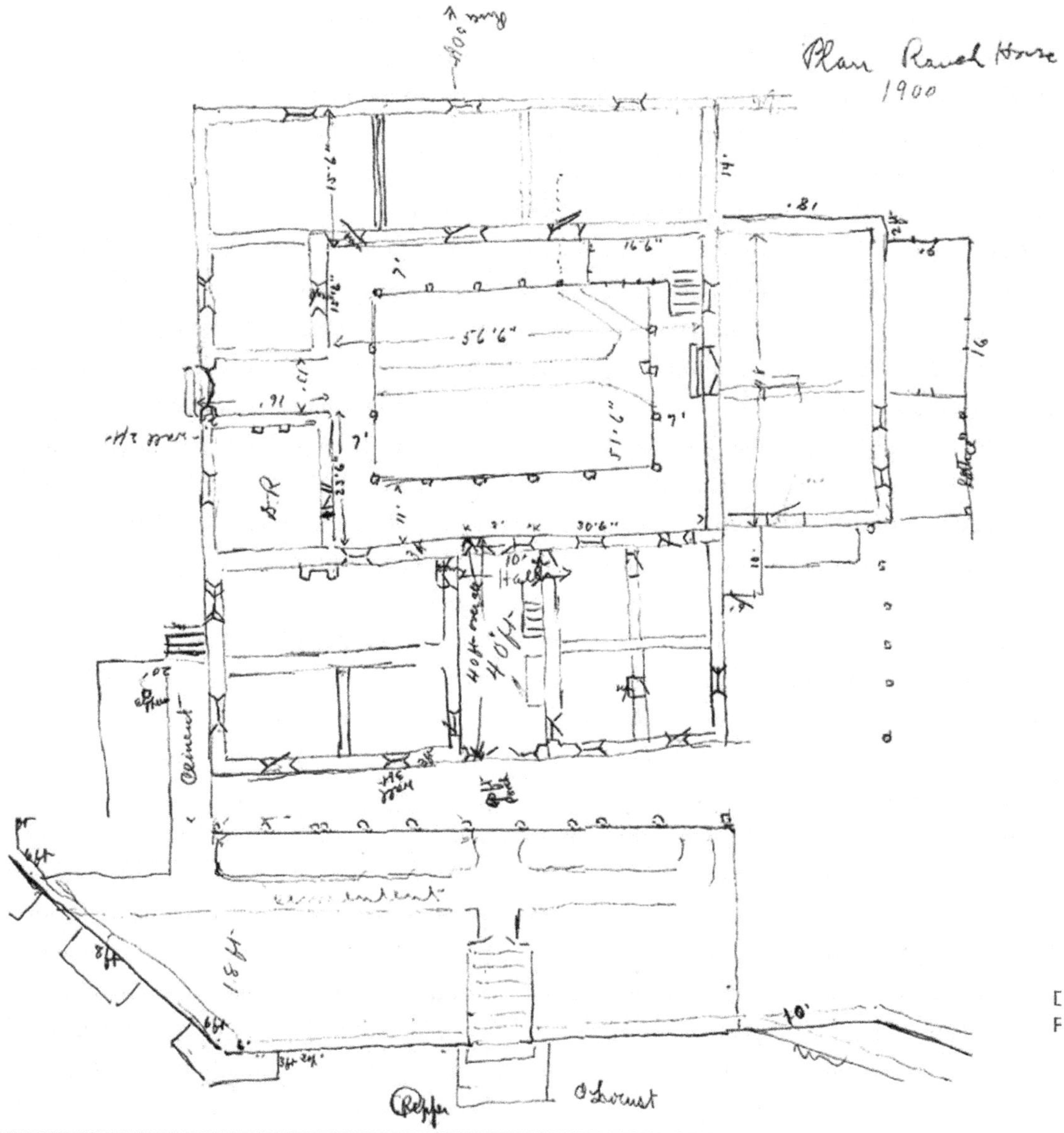

John Jay Baumgartner (1865–1946), a painter and architectural delineator who lived and worked in California, was related to the O'Neill family by marriage. A frequent visitor to the ranch houses at Santa Margarita and Las Flores, Baumgartner produced a hand-drawn floor plan and a scale model of the Santa Margarita Ranch House, ca. 1900.
Collection of Jerome Baumgartner Jr., MCB Camp Pendleton Archives

Jerome O'Neill. A major flood on 30 January 1916 precipitated significant updates to the Santa Margarita Ranch House. The natural disaster washed out the railroad bed and damaged the bunkhouse, shops, barns, and corral. The Ranch House, perched on a small knoll, fortunately did not sustain significant damage. After the floodwaters subsided, Jerome O'Neill undertook major construction to repair the damaged outbuildings and to complete an extensive remodeling of the Ranch House and bunkhouse.

Photographs from this period indicate that O'Neill made substantial additions at the northeast, southeast, and southwest corners of the house. Each expansion is characterized by a different construction scheme, although all have many features in common with respect to concrete detailing, millwork, and finishes. Whereas the southeast corner has a reinforced concrete frame with hollow clay infill, the main level addition at the northeast corner is constructed entirely of structural hollow clay tile. Additions to the northeast and southeast are two stories, with basements accessed at ground level below main level rooms. The addition at the southwest corner infills the space once occupied by the hallway on the south end of the main block.

Because the old roof had begun to leak by 1915, O'Neill reframed and replaced the roof, adapting roof lines to incorporate new additions. As part of this update, the stairs to the attic were removed and the upper floor ceased to be used as a living space. The installation of interior doorways established new traffic patterns throughout the house, as it became possible for the first time to pass from room to room without accessing the corredors of the courtyard and patio. With this reconfiguration of the house, the courtyard entrance on the north façade came to serve as the primary entrance to the home. The first bathtub also was reportedly installed in 1916, although the zinc tub had to be filled and emptied by hand, as indoor plumbing was not installed until approximately 1928. The house was equipped with its first electrical system at this time as well.

Jerome O'Neill oversaw the cattle ranching enterprise until his death in 1926, at which time the estate went to an assortment of heirs in both families. The property lacked a manager with the interest and expertise required to remain in the agricultural business, so the vast ranch was divided between the O'Neills, Floods, and Baumgartners (who had married into the O'Neill family) in 1938. The Flood descendants assumed the southern portion, and the northern portion split into two parts going to the O'Neill and Baumgartner families, respectively.

With the Marine Corps' purchase of the ranch in 1942, the Santa Margarita Ranch House became an enormous construction site. Working around the clock, laborers developed the landscape to provide airstrips, administration

At the entryway to the courtyard on the north side of the building, the doors were replaced with a more decorative set to reflect the increasing importance of this entryway in the reconfigured house.
Photo by the author

A new corredor was constructed in 1916 to create a covered walkway along the remodeled southwest façade. In addition, many new doors were added to connect interior rooms, thereby eliminating the need to use the corredors to move from room to room.
Historic American Buildings Survey, Library of Congress Prints and Photographs Division, courtesy of Henry F. Withey

buildings, and housing for the new military base. Alterations to the Ranch House complex took immediate effect. Numerous buildings associated with the ranch operation were demolished to make way for new construction necessary to the development of the base. By September 1942, only the adobe buildings—the Ranch House, bunkhouse, blacksmith shop, and outhouse—had not been demolished.

Base architects put forward two plans for redevelopment and reuse of the Santa Margarita Ranch House. The first plan, devised by Major General Clayton B. Vogel, proposed substantial alteration of the existing building with the addition of a wing enclosing a second courtyard. The

Few changes occurred on the property between Jerome O'Neill's 1916 remodeling project and the Marine Corps acquisition of the Santa Margarita Ranch House in 1942. With the transformation of the ranch to a military base during World War II, the Marine Corps razed many of the outbuildings surrounding the Ranch House.
Progress Edition, *1943, MCB Camp Pendleton Archives*

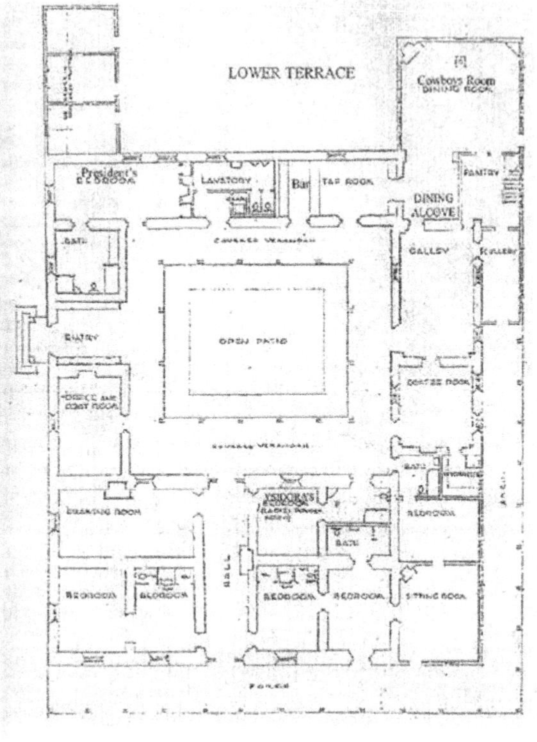

Plan of the Santa Margarita Ranch House, 1944.
MCB Camp Pendleton Archives

second plan, developed by California architects Myron Hunt and Harold C. Chambers, called for several minor changes directed to converting the old Ranch House into an officers' club. Detailed in a plan drawn by the architects in 1942, the proposed modifications focused on the south and east wings of the building and consisted of the addition of an arched opening between the dining room and tap room, a reconfigured doorway featuring *banco*, or bench, seating between the tap room and vestibule, and the installation of new casework. As part of this remodel, the architects designated a large kitchen with a scullery and pantry, from which the dining room could be directly accessed, as well as a coffee room with its own kitchenette. Hunt and Chambers further incorporated a lavatory and bathroom, which could be accessed from corredors on the east and south of the courtyard.

The Marine Corps secured funding and adopted the Hunt/Chambers plan to refurbish and adapt the Ranch House to suit its new purpose; however, when word of the extensive renovation leaked in August 1942, the California Historical Society intervened with an appeal to General Joseph C. Fegan to preserve the historical integrity of the Santa Margarita Ranch House complex. President Franklin D. Roosevelt also weighed in on the planned changes. While touring the Ranch House during his visit to base for the dedication ceremony on 25 September 1942, Roosevelt expressed his admiration for the historical structure and encouraged Brigadier General Fegan to adopt a policy of preservation and restoration. According to Fegan, the president instructed "that all restoration must proceed along historical lines; that as much of the old furniture as possible should be retained and re-procured; that the gardens, bunk house [*sic*] and winery should

Gen Joseph C. Fegan commissioned light fixtures for the corredors, main hall, and formal living spaces and began to assemble a collection of antiques to use as furnishings, such as the Empire-style sofa and armchairs in the *sala*, or reception room, pictured above. *Photo by the author*

be restored, and that the landscaping be carried out at points indicated by him."[10]

Under the careful watch of California Governor Culbert L. Olson and the California Historical Society, the Marine Corps restored and modernized the Santa Margarita Ranch House. As part of this construction, the southeast corner of the house was slightly reconfigured to support the officers' club function. Between 1942 and 1946, a bar was installed in the room adjacent to the large dining room. Both the kitchen and

[10] Joseph Fegan, letter dated February 1943, as quoted in Milford Wayne Donaldson, *Santa Margarita Ranch: United States Marine Corps Base Camp Joseph H. Pendleton* (San Diego, CA: Office of Historic Preservation, 1998), 36.

The Santa Margarita Ranch House served briefly as an officers' club in the 1940s. The meat hooks on the wall of the barroom recall its prior iteration as a cattle ranch, when the floor was slanted to drain blood dripping from freshly cut meat.
Historical Photograph Collection, MCB Camp Pendleton Archives

a buggy shed, which had previously not been attached to the house, were incorporated into the main structure. New doorways created a foyer, while the adobe wall on the north side of the main block was partially demolished, and a cast-in-place concrete wall was installed to stabilize the remaining adobe. Modern heating, plumbing, and electrical systems also were installed during this period. The arcaded walk along the south wall of the northeast corner was added sometime after the Marine Corps acquired the property as well.

The arbor entrance of the Ranch House has a bell that was once used at the Las Flores estancia. Cast in 1828 and brought to the small mission by Spanish priests from Mission San Luis Rey, the bell was donated to the Marine Corps by officials of the railway station at San Juan Capistrano, where it had hung since 1887. In 1943, the general manager of the Santa Fe Railway's Coast Lines, E. E. McCarty, presented the bell to Camp Pendleton.
Photo by the author

In addition, the bunkhouse underwent repairs and the winery was converted into a chapel as part of this modernization and restoration effort. Updated for use as quarters for noncommissioned officers, the bunkhouse remodel included the installation of a new tile roof. In 1949, the wooden corredor posts were replaced with concrete columns. In 1965, the Marine Corps established a museum in the building. The winery, meanwhile, had already been converted into a blacksmith shop during the O'Neill stewardship. The Marine Corps repaired the struc-

MURALS

The first commanding general to reside in the house was Major General Graves B. Erskine, who moved in with his family in July 1947. His wife, Peg, was responsible for two murals in the house. She painted the first herself. The second, painted by Master Sergeant Russ Vickers during World War II when the building served as an officers' club, nostalgically depicted an Old West scene of San Francisco's Barbary Coast during the mid-nineteenth century. A later resident of the Ranch House objected to the "lady of the evening" who was "clearly advertising her wares" in the dining room mural. Although the vignette was modified to depict the offending window with closed shades, the mural was subsequently covered in target paper and then plastered. In 1962, Mrs. John C. Munn uncovered Vicker's mural, which remains visible today in the vestibule near the southeast corner of the house.

Mural of the Barbary Coast of San Francisco by MSgt Russ Vickers. Commissioned by Graves B. Erskine's wife, the mural was considered too risqué by a later resident of the Ranch House. Vickers, a Marine combat artist during the Korean War, later worked as a technical illustrator for the aerospace industry in Los Angeles. Painting Western scenes in his spare time, he developed a naturalistic style. Several of his paintings appear in the film *Chisum* (1970) starring John Wayne.
Photo by the author

ture for use as a chapel for the Marine Corps Women Reserves in 1943. They repaired adobe walls and added a vestry, arcaded corredor, and bell tower framed in wood and stucco to emulate adobe. The funding for the latter project came primarily from private donations, including money given by the O'Neill family, General Fegan, and the widow of James L. Flood.

The Marine Corps ceased using the Santa Margarita Ranch House as an officers' club in 1946. As the building became quarters for the commanding general at Camp Pendleton the following year, the historical structure underwent further adaptation to suit this new purpose. Alterations included the installation of a central heating system, updating the electrical system, renovating existing bathroom facilities, and converting the kitchenette in the south wing to an entry foyer. New windows and doors accompanied these changes, as did the installation of casework in closets. To accommodate the new central heating system, the scullery was converted into a mechanical room.

Additional remodeling to update the kitchen and all of the bathrooms in the general's quarters has occurred since the 1950s. In 1951, then-resident of the Ranch House General Clifton B. Cates ordered that "no modification of this building, its immediate surroundings, or out-buildings, shall be made which might possibly affect its architecture or vary its character."[11] Since then, most of the changes fall under maintenance or have concentrated in portions of the building erected in 1916 to leave the adobe portions historically intact. It has undergone building systems upgrades, and facilities have generally been additive, preserving the significant historical fabric of the structure. Following a flood in 1993, during which the chapel sustained significant damage but the Ranch House was only minimally impacted, the Marine Corps constructed a levee to the north and east of the Santa Margarita Ranch House complex.

Today, the Santa Margarita Ranch House, Bunkhouse Museum, and chapel form the Ranch House National Historic Site. A total of 36 base commanders lived in the house between 1947 and October 2007. All had distinguished careers. Currently unoccupied, the Ranch House is being maintained in its present state while awaiting funding for an extensive upgrade project in 2018 to comply with earthquake stabilization requirements. Pending this essential preservation work, the historic Santa Margarita Ranch House will continue to serve the Marine Corps and surrounding community as a multiuse gathering place equipped with meeting rooms, exhibition space, and comfortable lodging for Camp Pendleton's distinguished guests.

As the oldest building at the Santa Margarita Ranch House complex, the Ranch House chapel operated as a winery, a blacksmith's shop, and as a home for the Pico family during the course of its long history. When the first women Marine reservists arrived at Camp Pendleton in 1943, the structure was renovated as a chapel and opened primarily for their use. Today, the Ranch House chapel holds weekly services and hosts weddings and other special events.
Official U.S. Marine Corps photo

[11] Witty and Morgan, *Marines of the Margarita*, 55.

Mrs. Sandi Boomer, wife of LtGen Walter E. Boomer, oversaw a redesign of the kitchen during her residency in the Ranch House.
Photo by the author

The fountain in the courtyard was donated by the only Marine Corps reservist to reside in the building, MajGen Hugh W. Hardy.
Official U.S. Marine Corps photo, courtesy of Cpl Shannon E. McMillan

Southern entrance to the Santa Margarita Ranch House as it appears today.
Photo by the author

ARCHITECTURAL FEATURES OF THE LAS FLORES ADOBE NATIONAL HISTORIC LANDMARK

Largely undisturbed by subsequent development of the Marine Corps base, the Las Flores Adobe site remains one of only a few in the region that represents a microcosm of Southern California history. Archaeological and historical evidence in the vicinity indicates nearly 2,000 years of occupation by Native American tribes prior to the arrival of Franciscan missionaries in the eighteenth century. When exactly Spanish priests from Mission San Luis Rey began to use the lands at Las Flores for farming and raising sheep remains unknown; however, the missionaries were established enough by April 1810 to complain to Comandante Francisco Ruiz of San Diego that Spanish troops were occupying one of their pastures.

After Mexico gained independence from Spain, the old mission buildings were the site of the Battle of Las Flores. In 1838, rival politicians Carlos Antonio Carrillo and Juan Bautista Alvarado clashed over the appointment of a provisional governor of California. Both mounted armies of approximately 200 men, and Carrillo retired to the Las Flores estancia or outpost where he used an old adobe building as a barracks and a corral as a fort. A standoff ensued, and only after cannon shots were fired did the two leaders and their representatives agree to a détente. On 23 April 1838, the two sides signed a treaty granting the power of government to Alvarado.

During the Mexican period, a decree of secularization broke up the large land holdings of Spanish missions, including Mission San Luis Rey. The Mexican government deeded 20,000 acres of land to Native Americans residing near the former Las Flores estancia. Belonging communally to 32 Luiseño families, the site became a *pueblo libre* or free village called Ushmai. Brothers Andrés and Pio Pico, who owned the adjacent Santa Margarita ranch, purchased the parcel from the Luiseños in 1844.[1]

Following the Mexican-American War and the transfer of California to U.S. governance, the Picos sold the Rancho Santa Margarita y Las Flores to their brother-in-law, John Forster. In 1867, Forster gave a parcel of the ranch at Las Flores to his son, Marco A. Forster, to construct Las Flores Adobe as a home for his family.[2] The structure is one of the few authentic nineteenth-century, two-story adobe ranch houses combining the Hacienda and Monterey Colonial styles, which are unique to California. Marked by arcaded silhouettes, spindled window grills, and heavy wooden colonial doors, the Hacienda style celebrates natural materials of the Southwest and relies on decorative tiling and an earthy color palette. Used primarily for residential properties, the Monterey Colonial

Portrait of Marco A. Forster (1893–1904). John Forster's sons took an active role in the daily operations of the ranch. His oldest son, Marco, led the rodeos and managed the farm at Las Flores. Chico and Juan Jr. managed the vaqueros leading the cattle drives, while Thomas handled the sheep business. In 1867, Marco Forster began construction on the Ranch House at Las Flores, where he would reside with his wife and children until 1882.
Ancestry.com

[1] This transaction was not without controversy. In February 1840, Pio Pico reported that he had turned over the property at Las Flores to a native man, Jose Manuel, but by May of that year, Father Francisco Ibarra of San Luis Rey wrote to another priest that Las Flores had become a *seraglio* for the Pico family. The indigenous inhabitants denounced Pio Pico and asked the government to intervene to force him to vacate any dwellings and remove his cattle from the territory. Following American occupation and subsequent transition of governmental power, the Pico brothers filed a petition with the U.S. Land Commission claiming possession of the land and buildings at Las Flores. On 24 April 1855, the board confirmed ownership rights for the Pico brothers, and after a series of appeals during the next decade, the Rancho Santa Margarita y Las Flores was legally established as belonging to the Pico family.

[2] Historical lore maintains that John Forster gave the Las Flores property to his son as a wedding gift; however, Marco was already with his wife when he began building their home in 1867. In addition, their first child was born four years earlier, in 1863, when Marco was 23 years old. Whether the elder Forster allowed Marco to live and build a home on his estate due to local custom or convenience, the arrangement likely was not a wedding gift. The author wishes to thank Faye Jonason, museum director at MCB Camp Pendleton, for her insights and expertise on this topic.

An estancia, or mission outpost, was established near the coast at the mouth of Las Flores Creek sometime before 1827. Constructed of adobe, the buildings consisted of a residence, granaries, and a chapel arranged around a central courtyard.
Sketch by H. M. T. Powell, March 1850, MCB Camp Pendleton Archives

Las Flores Adobe National Historic Landmark.
Official U.S. Marine Corps photo

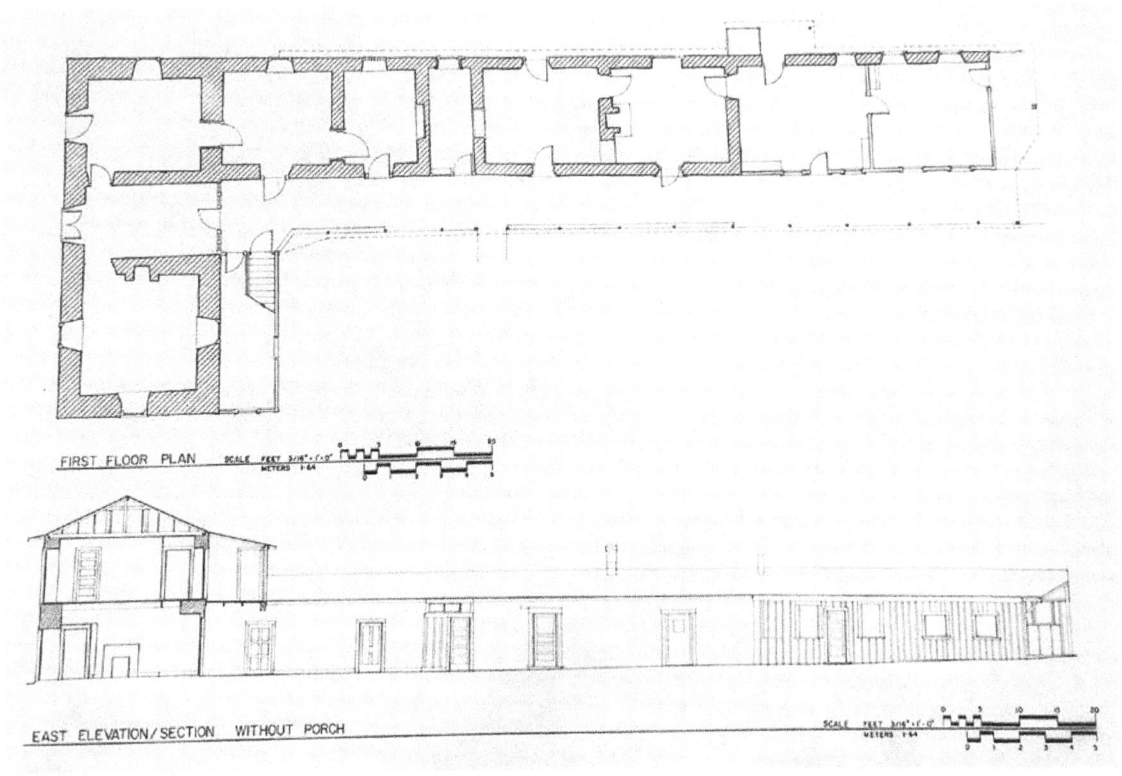

Plan of first floor from east elevation, including detail of foundation and walls.
Plans by Jake Barrow, National Park Service (Ret)

style emerged from early Californians' desire to merge local Spanish influences with Colonial designs from the East Coast. Developed in the early part of the eighteenth century, the Monterey style grew in popularity between 1930 and 1950. Architecture in this style typically features two stories with continuous surrounding porches on both levels, a hipped roof, and adobe walls.

Of the numerous buildings that originally comprised the complex, only Las Flores Adobe and the Carriage House survive. The most prominent component of Las Flores Adobe is the Monterey Colonial block, a two-story, one-room-deep, hipped-roof structure. Characterized

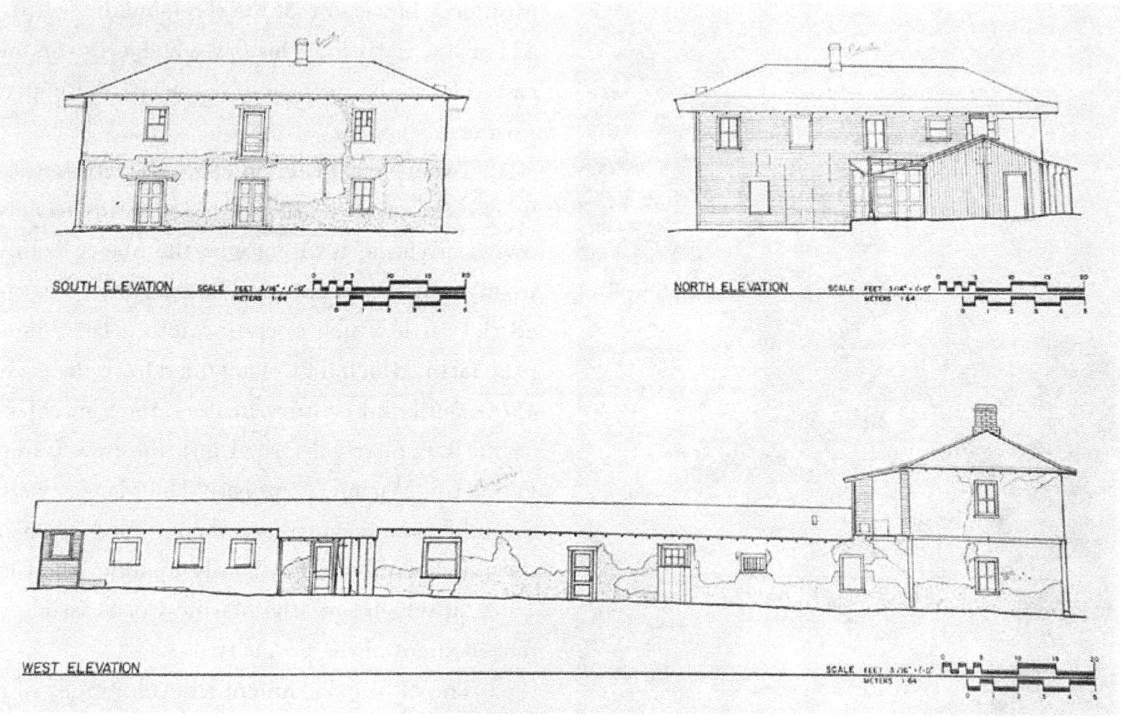

Plan of first and second floors from west, south, and north elevations.
Plans by Jake Barrow, National Park Service (Ret)

by thick adobe walls that taper as they ascend, the building is laid out around a central hallway, or *zaguan*, that opens onto the porch of the Hacienda wing. High-style porches originally adorned the front and rear, but were later replaced by a porch wrapping the entirety of the block; a covered balcony also was added on the second level facing the Pacific Ocean. The Hacienda block is the long, single-story portion connecting the Monterey block of Las Flores Adobe and the Carriage House. Mostly utilitarian, the low, gable-roofed structure is one room deep and features a continuous shed-roof corredor, or covered porch. The Carriage House, also gable-roofed, is located at the north end of the Hacienda block and oriented parallel to the

The Magee family resided at Las Flores Adobe ranch house between 1888 and 1968, at which time the Marine Corps assumed management of the property. During their occupancy, the Magees converted the cattle ranch operation into a large lima bean farm, which they maintained from the early 1900s until mid-century. This photograph, taken in 1938, shows the wedding announcement dinner for Mrs. Jane Magee Stauss.
MCB Camp Pendleton Archives, courtesy of Bob Stauss

Monterey block so that the three buildings form a U-shape. Early in its history, a hedgerow on the east side of the property completed the courtyard enclosure.

Twenty years later, in 1888, Las Flores Adobe passed out of the Pico/Forster family to new owners, who in turn engaged the Magee family to manage the property. The Magees converted the cattle ranch operation into a large lima bean farm, which they maintained from the early 1900s until mid-century. In 1942, the Santa Margarita Ranch was absorbed into the new Camp Pendleton Marine Corps base. The Magees were given lifetime occupancy and continued living at the ranch until the last family member died in 1968, at which time the Marine Corps assumed management of the property.

The U.S. government's development of a military base on the former ranch land altered its primary usage and removed maintenance incentives and proprietary interests from local residents, resulting in very expedient and negative alterations to the site. Many of the adobe buildings were destroyed over time, as the general impression incorrectly assumed that they were structurally unsuitable for contemporary use. By 1968, the Las Flores Adobe and surrounding buildings were in an advanced state of disrepair. Threatened with demolition due

Las Flores Adobe, ca. 1935. Renovations made to the ranch house around 1918 enclosed the two-story porch and extended the corredor the entire length of the Hacienda block.
MCB Camp Pendleton Archives, courtesy of Bob Stauss

to its poor condition, the property inspired local citizens to mobilize in a preservation campaign. In 1968, Las Flores Adobe was placed on the National Register of Historic Places and designated a National Historic Landmark.[3]

As the buildings at Las Flores Adobe had not been used since the last resident departed in the 1960s, the Marine Corps sought a tenant to help maintain the house and surrounding property. In 1974, the Marines leased Las Flores to

[3] The National Historic Landmark designation is made by the secretary of the interior and is reserved for historical sites that possess exceptional value in illustrating the heritage of the United States. This designation is the highest level of federal recognition available for cultural resources. Whereas the nation's National Register of Historic Places includes more than 85,000 sites, the National Historic Landmark program recognizes only about 2,500 buildings, objects, or sites as having cultural heritage of national importance.

The Las Flores Adobe complex from the southeast, ca. 2008.
MCB Camp Pendleton Archives

the Boy Scouts of America, who carried out extensive modifications to the Carriage House. The remodeling project involved the complete removal of the structure's roof covering, large doors, and shed additions. The Boy Scouts also installed a concrete beam atop the adobe walls, increasing their height by approximately eight inches, and reframed the roof using modern materials. A concrete *contra pared*, or skirting wall, was added against the foundation stonework and along the lowest courses of adobe on the building perimeter, and a concrete floor was poured on the interior. Both interior and exterior walls were covered with a hard, relatively impermeable Portland cement plaster.

Deterioration of Las Flores Adobe and the Carriage House nevertheless continued as minimal resources were available, and by the 1990s, the site was closed for safety reasons. Recognizing the value and declining condition of Las

Aerial shot of Las Flores Adobe historic site.
Camp Pendleton Historical Society

Flores Adobe, the Marine Corps registered the property with the National Trust for Historic Preservation's most endangered structures list.

When funds finally became available in 1999, the Marine Corps contracted the National Park Service to assist with the stabilization and restoration effort at Las Flores Adobe. Preservation work included restructuring the floor foundation, removing vegetation, improving drainage and ventilation, applying protective lime plaster, and restoring the roof. Located in a seismic zone—just several kilometers east of a major fault running along the coastline of Southern California—Las Flores Adobe also underwent seismic stabilization. Whereas previous approaches to earthquake stabilization had required major demolition to the historic fabric of the structure, the team at Las Flores installed minimally invasive systems using rods, steel strapping, grouted pins, and plywood sheer panels. Recent renovations also restored the lost two-story porch, which completely surrounds the Monterey section, to better protect the adobe walls.

Renovations to the Carriage House implemented in 2004 included the replacement of an asphalt roof covering with cedar shingles, installation of a seismic retrofit system, repair of decayed woodwork, and conservation of earthen and lime finishes. This project also entailed the reinstatement of a connection between the Hacienda block and Carriage House, which had to be modified due to the 1974 change in elevation and pitch of the Carriage House roof. Because the design could not replicate the original configuration, the Las Flores Adobe roof extends to the Carriage House only to the extent re-

Exterior view of Las Flores Adobe showing how the change of elevation enacted in 1974 resulted in the modern roof extension abutting the wall rather than the eaves of the Carriage House.
MCB Camp Pendleton Archives

quired to weather-protect the walkway between the buildings.

Following the success of this project, the Marine Corps has maintained its collaboration with staff and graduate students from the University of Vermont Historic Preservation Program to continue its preservation and rehabilitation efforts at Las Flores Adobe.

Restoration work on Las Flores Adobe began in 2002 by the University of Vermont with the support of the National Park Service and the U.S. Marine Corps. Surviving plaster was cleaned, and wholesale plaster loss was replaced in kind. New layers of limewash were applied where needed. The National Park Service completed structural stabilization of the main house in 2004.
MCB Camp Pendleton Archives

Las Flores Adobe National Historic Landmark as it appears today.
Official National Park Service photo

APPENDIX A
TIMELINE

1769: Capt Gaspar de Portolá leads an expeditionary force to the location now known as Camp Pendleton. They baptize the land "Santa Margarita" in honor of Saint Margaret of Antioch, the fourth-century martyr on whose feast day (20 July) the expedition arrives.

1821: The Mexican War of Independence ends Spanish rule in the territory of New Spain, which results in Mexico taking possession of California.

1827: An inventory by Father Antonio Peyri notes the existence of adobe structures at the current site of the Santa Margarita Ranch House.

1838: Rival politicians Carlos Antonio Carrillo and Juan Bautista Alvarado faced off in the Battle of Las Flores to determine the governorship of California. Following a military standoff at the site of the Las Flores estancia, the two sides signed a treaty granting the power of government to Alvarado.

1841: Brothers Pio and Andrés Pico receive a land grant in Alta California for a parcel now known as Rancho Santa Margarita y Las Flores from Mexican Governor Juan Bautista Alvarado.

1846: The United States declares war on Mexico. A detachment of Marines from the sloop-of-war *Cyane* make landfall and raise American and Marine Corps flags in the Old Town Plaza of San Diego. Later that year, a joint force of U.S. Army dragoons, Marines, and sailors challenge Californio lancers, led by Andrés Pico, at the Battle of San Pasqual.

1848: The United States acquires California as part of the Treaty of Guadalupe Hidalgo.

1864: The Pico brothers sell their shares of

Rancho Santa Margarita y Las Flores to their English brother-in-law, John Forster, to pay off their debts.

1867: Marco Forster, the eldest son of John Forster, constructs Las Flores Adobe as a residence for his family.

1882: John Forster dies and leaves the Santa Margarita Ranch House and property to his heirs, who are forced to sell the land following a series of droughts and a new fence law. Wealthy cattleman James C. Flood and his manager, Richard O'Neill, acquire the ranch.

1888: The Magee family leases the Las Flores Adobe complex.

1910 Richard O'Neill dies. His eldest son, Jerome, inherits his father's interest in the property and takes over his management duties.

1911: President William H. Taft dispatches the 4th Provisional Marine Regiment to San Diego in response to the Mexican Revolution. The Marines establish Camp Thomas on North Island.

1914: Called back to San Diego in response to continued Mexican political and civil disorder, Col Joseph H. Pendleton leads the 4th Marine Regiment to the Gulf of California as a show of force.

1915: Col Pendleton and Capt Charles H. Lyman, commanding officer and adjutant 4th Regiment, oversee an encampment of Marines at the Panama-California Exposition in San Diego's Balboa Park.

1916: Overflow of the Santa Margarita River causes extensive flooding, resulting in a major repair and renovation project at the Santa Margarita Ranch House complex under the stewardship of Jerome O'Neill.

1921: The Marines move to Dutch Flats, site of the Marine Advanced Base.

1923: The Marine Corps recruit depot for the western United States is transferred from the Marine Barracks, Mare Island Naval Shipyard, California, to the new post in San Diego.

1924: The recruit depot at San Diego receives the official designation of Marine Corps Base, Naval Operating Base, San Diego, on 1 March 1924.

1926: Jerome O'Neill dies.

1927: Marine Corps units bound for China assemble and embark at San Diego in the largest operation of this kind at Marine Corps Base (MCB) San Diego prior to World War II.

1935: Established in 1933, Headquarters, Fleet Marine Force, transfers from MCB Quantico to MCB San Diego.

1941: The 1st Marine Brigade becomes the 1st Marine Division in February. The United States declares war on Japan and Germany in response to the attack on Pearl Harbor.

1942: The U.S. Navy acquires Rancho Santa Margarita y Las Flores from the Flood and O'Neill families and commissions architect Myron H. Hunt and his firm Hunt, Chambers, and Ellingwood to design several buildings on the new base, including the administrative building now used as 1st Marine Division Headquarters. At the suggestion of President Franklin D. Roosevelt, the historic Santa Margarita Ranch House is preserved and used as quarters for distinguished visitors.

1943: Women Marine Reserves arrive at Camp Pendleton. The 4th Marine Division is activated at Camp Pendleton. Sgt Wayne Richards wins a contest to select an emblem for the base; the Marines adopted the O'Neill cattle brand—a hooked "T" atop an "O"—as Camp Pendleton's logo. Sgt Richards got the idea when he saw the brand on one of the ranch's earliest buildings: the blacksmith shop.

1944: Camp Pendleton is declared a "permanent" installation in October.

1946: Gen Alexander A. Vandegrift, Commandant of the Marine Corps, designates Camp Pendleton as the center of all West Coast Marine Corps activities and home to the 1st Marine Division.

1947: The 1st Marine Division arrives at Camp Pendleton after combat and occupation duty in the Pacific. MajGen Graves B. Erskine takes up residence at the Santa Margarita Ranch House, and his wife Peg commissions a mural of San Francisco's Barbary Coast by MSgt Russ Vickers. A later resident of the house considers the painting too racy and has it covered.

1950: The United States enters the Korean War. Reserve Marines arrive at Camp Pendleton and begin processing and training for subsequent deployment to Korea.

1953: Camp Pendleton is officially designated Marine Corps Base Camp Pendleton.

1960s: The 1st Division Headquarters building undergoes extensive modifications to the porch and other external elements. The 1st Marine Division deploys to Southeast Asia for involvement in the war in Vietnam.

1964: Congress passes a resolution granting President Lyndon B. Johnson free hand in Vietnam. At the Santa Margarita Ranch House, Mrs. J. C. Munn uncovers the mural by MSgt Vickers.

1966: The 4th Marine Division Reserves, Fleet Marine Force, is reactivated at Camp Pendleton.

1968: Las Flores Adobe is designated a National Historic Landmark by the National Park Service.

1971: The I Marine Amphibious Force (I MAF)

is relocated to Camp Pendleton. The Santa Margarita Ranch House is listed on the National Register of Historic Places.

1975: Camp Pendleton serves as a staging area for refugees arriving from South Vietnam during Operation New Arrivals.

1980s: The 1st Division Headquarters building undergoes major renovations that alter its interior and exterior appearance.

1988: I MAF is redesignated as I Marine Expeditionary Force (I MEF).

1990s: Marines from I MEF deploy to Saudi Arabia in support of Operations Desert Shield and Desert Storm.

1991: The 1st Marine Division assists in relief efforts in Bangladesh and the Philippines.

1992: I MEF participates in Operation Restore Hope.

1993: Overflow of the Santa Margarita River causes significant flooding at Camp Pendleton, resulting in serious damage to the Santa Margarita Ranch House, chapel, and numerous other structures on base.

2000s: Las Flores Adobe undergoes an extensive renovation and stabilization project.

2002: I MEF units deploy to Kuwait in support of Operation Iraqi Freedom (OIF).

2003: The 1st Marine Division deploys by air and sea to link up with its advanced headquarters in Kuwait under I MEF in support of OIF. The 1st Marine Division conducts the longest ground march in Marine Corps history, attacking Baghdad alongside the U.S. Army.

2004: The 1st Marine Division relieves the Army's 82d Airborne Division in Iraq's Sunni Triangle. During OIF II, 1st MarDiv conducts counterinsurgency operations throughout al-Anbar Province.

2005: Marine Corps Installations-West (MCIWEST) is reactivated at Camp Pendleton.

2006: The 1st Marine Division deploys to Iraq as the ground combat element for I MEF.

2007: MajGen Michael R. Lehnert becomes the last Marine general officer to occupy the Santa Margarita Ranch House.

2010: Personnel and units of the 1st Marine Division deploy to Afghanistan to provide advisory support and maneuver elements in support of Operation Enduring Freedom.

2012: The 1st Marine Division deploys to Afghanistan to serve as the headquarters for Task Force Leatherneck, the ground combat element for Regional Command-Southwest. MCIWEST merges with MCB Camp Pendleton to form Marine Corps Installations West-Marine Corps Base Camp Pendleton.

APPENDIX B
U.S. MARINE CORPS GENERAL RESIDENTS, SANTA MARGARITA RANCH HOUSE, 1947–2007

July 1947–July 1950	MajGen Graves B. Erskine
July 1950–July 1951	MajGen Alfred H. Noble
July 1951–May 1953	MajGen Oliver P. Smith
May–August 1953	MajGen Robert H. Pepper
November 1953–April 1955	MajGen John T. Selden
April 1955–July 1957	MajGen George F. Good Jr.
July 1957–July 1959	MajGen Reginald H. Ridgely
November 1959–March 1961	MajGen Alan Shapley*
June 1962–April 1963	MajGen James M. Masters Sr.
May 1963–June 1964	MajGen John C. Munn
July 1964–April 1967	MajGen Robert E. Cushman Jr.
March 1967–June 1968	MajGen Lewis J. Fields
July 1968–August 1970	MajGen Donn J. Robertson
August 1970–December 1971	MajGen George S. Bowman Jr.
January 1972–November 1973	MajGen Herman Poggemeyer Jr.
November 1973–August 1974	MajGen Robert L. Nichols
August 1974–April 1975	BGen Paul G. Graham
June 1975–June 1978	MajGen Carl W. Hoffman

* BGen Raymond L. Murray served as base commanding general from March 1961 to June 1962; however, he never resided in the Ranch House as a result of orders from then-Commandant Gen David M. Shoup, who instructed Murray to await a new assignment.

July 1978–June 1980	MajGen Stephen G. Olmstead
June 1980–August 1980	MajGen Hugh W. Hardy (shortest stay at 52 days)
August 1980–August 1983	MajGen Kenneth L. Robinson
August 1983–July 1984	MajGen Anthony Lukeman
July 1984–June 1987	MajGen Robert E. Haebel
July 1987–April 1988	BGen Matthew T. Cooper
June 1988–June 1990	BGen Richard H. Huckaby
August 1990–September 1991	LtGen Walter E. Boomer
September 1991–June 1993	LtGen Robert B. Johnston
July 1993–May 1994	LtGen George R. Christmas
June 1994–April 1996	LtGen Anthony C. Zinni
September 1996–August 1998	MajGen Claude W. Reinke
August 1998–July 2000	LtGen Bruce B. Knutson Jr.
July 2000–November 2002	LtGen Michael W. Hagee
November 2002–September 2004	LtGen James T. Conway
October 2004–August 2006	LtGen John F. Sattler
August 2006–September 2007	MajGen Michael R. Lehnert

APPENDIX C
COMMANDERS OF MARINE CORPS ACTIVITIES, SAN DIEGO, CALIFORNIA, 1914–42

MARINE BARRACKS, U.S. NAVAL STATION, SAN DIEGO, CALIFORNIA, 1914–24

December 1914–June 1915	Maj William N. McKelvey
July 1915–January 1916	Capt Ellis B. Miller
January–February 1916	Maj William N. McKelvey
February–June 1916	Col Joseph H. Pendleton
June–September 1916	1stLt Seldon B. Kennedy*
September–December 1916	1stLt Earl C. Long*
December 1916–January 1917	Capt Thomas C. Turner
January 1917	1stLt Earl C. Long
January–October 1917	Maj Thomas C. Turner*
October 1917–October 1918	LtCol Carl Gamborg-Andresen
October 1918–April 1919	Maj David M. Randall
April 1919–July 1921	Col John F. McGill
July–October 1921	None designated
October 1921–January 1922	LtCol James McE. Huey
January–March 1922	Maj Eugene P. Fortson
March 1922–February 1924	LtCol Giles Bishop, Jr.

* Individuals marked with an asterisk received a promotion during their tenure as commander.

MARINE CORPS BASE SAN DIEGO, CALIFORNIA, 1924–42

March 1924	MajGen Joseph H. Pendleton
April–May 1924	Col James McE. Huey
May–June 1924	MajGen Joseph H. Pendleton
June 1924–April 1925	Col John T. Myers
April–May 1925	LtCol William H. Pritchett
May–October 1925	Col John T. Myers
November 1925–February 1926	Col Alexander S. Williams
February–March 1926	BGen Smedley D. Butler
March–May 1926	LtCol William H. Pritchett
May 1926–March 1927	BGen Smedley D. Butler
March 1927	Maj Benjamin A. Moeller
April–August 1927	LtCol William H. Pritchett
August 1927–June 1928	Col Charles H. Lyman
June–July 1928	Maj Benjamin A. Moeller
July 1928–April 1929	BGen Dion Williams
April 1929–January 1930	Col Harry R. Lay
January–December 1930	BGen Robert H. Dunlap
December 1930–November 1931	BGen John H. Russell
November–December 1931	Col Charles H. Lyman
December 1931–January 1932	BGen Frederick L. Bradman
January–February 1932	Col Harry R. Lay
February 1932–December 1933	BGen Frederick L. Bradman
December 1933–January 1934	Col Rush R. Wallace
January–March 1934	BGen Frederick L. Bradman
March–June 1934	Col Rush R. Wallace
June 1934–April 1935	BGen Frederick L. Bradman
May 1935	Col Rush R. Wallace
May 1935–January 1937	BGen Douglas C. McDougal
January–February 1937	Col Alley D. Rorex
February–May 1937	BGen Douglas C. McDougal

May 1937–March 1938	MajGen Louis Mac. Little
March–April 1938	LtCol Harry L. Smith
April 1938–August 1939	MajGen Louis Mac. Little
August 1939	BGen Clayton B. Vogel
September 1939	BGen Richard P. Williams
September 1939–December 1941	BGen William P. Upshur
December 1941–March 1942	Col William H. Rupertus
March–April 1942	Col Matthew H. Kingman
April–August 1942	Col James L. Underhill

APPENDIX D
BASE COMMANDING GENERALS, CAMP PENDLETON, 1942–PRESENT

MARINE CORPS BASE CAMP PENDLETON COMMANDING GENERALS

August 1942–May 1944	MajGen Joseph C. Fegan
June 1944–July 1945	MajGen Charles F. B. Price
July 1945–May 1946	LtGen Holland M. Smith
May 1946–June 1947	LtGen Harry Schmidt
June–July 1947	BGen John T. Walker
July 1947–July 1950	MajGen Graves B. Erskine
July 1950–June 1951	MajGen Alfred H. Noble
June 1951–May 1953	MajGen Oliver P. Smith
May–August 1953	MajGen Robert H. Pepper
August–October 1953	BGen William J. Scheyer
October–November 1953	MajGen James P. Riseley
November 1953–April 1955	MajGen John T. Seldon
April 1955–July 1957	MajGen George F. Good Jr.
July 1957–October 1959	MajGen Reginald H. Ridgely Jr.
November 1959–March 1961	MajGen Alan Shapley
March 1961–June 1962	BGen Raymond L. Murray
June 1962–April 1963	MajGen James M. Masters Sr.

May 1963–June 1964	MajGen John C. Munn
July 1964–March 1967	MajGen Robert E. Cushman Jr.
March 1967–June 1968	MajGen Lewis J. Fields
June–July 1968	MajGen Wood B. Kyle
July 1968–August 1970	MajGen Donn J. Robertson
August 1970–December 1971	MajGen George S. Bowman Jr.
December 1971–November 1973	MajGen Herman Poggemeyer Jr.
November 1973–August 1974	MajGen Robert L. Nichols
May–June 1974	BGen Leonard E. Fribourg
August 1974–June 1975	BGen Paul G. Graham
June 1975–June 1978	MajGen Carl W. Hoffman
July 1978–June 1980	MajGen Stephen G. Olmstead
June–August 1980	MajGen Hugh W. Hardy
August 1980–August 1983	MajGen Kenneth L. Robinson
August 1983–July 1984	MajGen Anthony Lukeman
July 1984–June 1987	MajGen Robert E. Haebel
June 1987–April 1988	MajGen Matthew T. Cooper
April–June 1988	MajGen James J. McMonagle
June 1988–June 1990	BGen Richard H. Huckaby
June–August 1990	MajGen John P. Monahan
August–September 1990	LtGen Walter E. Boomer
September 1990–June 1991	BGen Michael I. Neil
June–September 1991	LtGen Walter E. Boomer
September 1991–June 1992	LtGen Robert B. Johnston
June 1992–May 1994	MajGen B. Don Lynch
May 1994–August 1998	MajGen Claude W. Reinke
August 1998–June 2001	MajGen Edward Hanlon Jr.
June 2001–June 2002	MajGen David F. Bice
June 2002–July 2004	MajGen William G. Bowdon
July 2004–April 2005	MajGen Timothy E. Donovan
April–October 2005	MajGen Michael R. Lehnert

October 2005–May 2006	Col John C. Coleman
May 2006–June 2009	Col James B. Seaton III
June 2009–March 2012	Col Nicholas F. Marano

MARINE CORPS INSTALLATIONS WEST (MCIWEST)* COMMANDING GENERALS

October 2005–September 2009	MGen Michael R. Lehnert
September 2009–November 2011	MajGen Anthony L. Jackson
November 2011–April 2012	BGen Vincent A. Coglianese

MCIWEST-MARINE CORPS BASE CAMP PENDLETON COMMANDING GENERALS

April 2012–August 2013	BGen Vincent A. Coglianese
August 2013–August 2014	BGen John W. Bullard
August 2014–July 2016	BGen Edward G. Banta
July 2016–present	BGen Kevin J. Killea

* MCIWEST was reactivated in 2005 and then MCIWEST and MCB Camp Pendleton were merged in 2012.

APPENDIX E
BASE SERGEANTS MAJOR, CAMP PENDLETON, 1956–PRESENT

MARINE CORPS BASE CAMP PENDLETON SERGEANTS MAJOR

November 1956–February 1960	SgtMaj Edwin L. Schwaner
March–December 1960	SgtMaj Milburn N. Zell
January 1961–September 1962	SgtMaj Otis B. Joyner
October 1962–October 1966	SgtMaj William J. Scheffer
October 1966–May 1968	SgtMaj Edgar R. Huff
May–October 1968	SgtMaj Warren W. McElliot
October 1968–April 1970	SgtMaj George E. Parker
May 1970–February 1972	SgtMaj Homer L. Akin Jr.
March–August 1972	SgtMaj Mike D. Mervosh
August 1972–September 1975	SgtMaj Henry M. Kajdacz
October 1975–October 1978	SgtMaj Daniel A. Misernay Sr.
November 1978–March 1980	SgtMaj L. C. Johnson
March–September 1980	SgtMaj Eleanor L. Judge
September 1980–September 1982	SgtMaj Donald L. Dolan
October 1982–November 1983	SgtMaj Lawrence C. Lookenbill
November 1983–January 1984	SgtMaj Dearl R. Fraze
February 1984–July 1985	SgtMaj Joe R. Rodriguez
July 1985–July 1986	SgtMaj Raymond T. Fitzhugh

July 1986–March 1987	SgtMaj Lee M. Bradley
March 1987–April 1990	SgtMaj Michael J. Baumhover
May 1990–September 1991	SgtMaj Harold A. Robinson
September 1991–November 1992	SgtMaj James I. L. Celestine
November 1992–June 1994	SgtMaj Mikel R. Beal
June 1994–September 1995	SgtMaj Chris A. Crawford
September 1995–November 1997	SgtMaj Richard W. Smith
November 1997–December 1998	SgtMaj Carlos R. Ramirez
December 1998–June 2000	SgtMaj Richard E. Jones
June 2000–August 2002	SgtMaj Michael G. Markiewicz
August 2002–April 2004	SgtMaj Jerome B. Price
April 2004–February 2006	SgtMaj Juan F. Sandoval Jr.
February–November 2006	SgtMaj Javier A. Nicholas
November 2006–January 2007	SgtMaj Wayne R. Bell
January–May 2007	SgtMaj Kenneth W. Jones
May 2007–March 2009	SgtMaj Jeffrey H. Dixon
March 2009–April 2012	SgtMaj Ramona D. Cook

MARINE CORPS INSTALLATIONS WEST (MCIWEST) SERGEANTS MAJOR

February 2006–January 2007	SgtMaj Wayne R. Bell
January 2007–April 2009	SgtMaj Barbara J. Titus-Tention
April 2009–June 2011	SgtMaj Jeffrey H. Dixon
June 2011–April 2012	SgtMaj Derrick Christovale Sr.

MCIWEST-MARINE CORPS BASE CAMP PENDLETON SERGEANTS MAJOR

April 2012–June 2013	SgtMaj Derrick Christovale Sr.
June 2013–May 2016	SgtMaj Scott R. Helms
May 2016–present	SgtMaj Julio E. Meza

APPENDIX F
COMMANDING GENERALS OF 1ST MARINE DIVISION, CAMP PENDLETON, 1941–PRESENT

February 1941–June 1941	MajGen Holland M. Smith
June 1941–March 1942	MajGen Philip H. Torrey
March 1942–July 1943	MajGen Alexander A. Vandegrift
July 1943–November 1944	MajGen William H. Rupertus
November 1944–August 1945	MajGen Pedro A. del Valle
August 1945–June 1946	MajGen DeWitt Peck
June–September 1946	MajGen Keller E. Rockey
September 1946–June 1947	MajGen Samuel L. Howard
June–July 1947	Col Alva B. Lasswell (Acting)
July 1947–July 1950	MajGen Graves B. Erskine
July 1950–February 1951	MajGen Oliver P. Smith
February–March 1951	BGen Lewis B. Puller (Acting)
March–April 1951	MajGen Oliver P. Smith
April 1951–January 1952	MajGen Gerald C. Thomas
January–August 1952	MajGen John T. Selden
August 1952–June 1953	MajGen Edwin A. Pollock
June 1953–May 1954	MajGen Randolph McC. Pate
May–July 1954	MajGen Robert H. Pepper

July 1954–January 1955	MajGen Robert E. Hogaboom
January 1955–August 1956	MajGen Merrill B. Twining
August 1956–June 1957	MajGen Robert O. Bare
July 1957–January 1958	MajGen David M. Shoup
January–February 1958	BGen Harvey C. Tschirgi
February 1958–October 1959	MajGen Edward W. Snedeker
October–November 1959	BGen Thomas F. Riley
November 1959–June 1961	MajGen Henry R. Paige
June–July 1961	BGen Frederick E. Leek
July 1961–June 1962	MajGen James M. Masters Sr.
June 1962–April 1963	MajGen Herman Nickerson Jr.
April 1963–July 1965	MajGen William T. Fairbourn
July–August 1965	BGen Edward H. Hurst
August 1965–September 1966	MajGen Lewis J. Fields
October 1966–May 1967	MajGen Herman Nickerson Jr.
June 1967–June 1968	MajGen Donn J. Robertson
June–December 1968	MajGen Carl A. Youngdale
December 1968–December 1969	MajGen Ormond R. Simpson
December 1969–April 1970	MajGen Edwin B. Wheeler
April 1970–April 1971	MajGen Charles F. Widdecke
April 1971–August 1972	MajGen Ross T. Dwyer Jr.
August 1972–April 1973	BGen Adolph G. Schwenk
May 1973–August 1974	MajGen Kenneth J. Houghton
August 1974–June 1975	BGen William L. McCulloch
June 1975–July 1976	MajGen Charles D. Mize
July 1976–July 1977	MajGen Edward A. Wilcox
July–August 1977	BGen Marc A. Moore
August 1977–August 1979	MajGen Charles G. Cooper
August 1979–July 1980	MajGen Francis X. Quinn
August 1980–August 1982	MajGen James L. Day
August 1982–June 1985	MajGen Ernest C. Cheatham
July 1985–July 1988	MajGen James J. McMonagle

July 1988–August 1990	MajGen John P. Monahan
August 1990–July 1992	MajGen James M. Myatt
July 1992–June 1994	MajGen Charles E. Wilhelm
June 1994–June 1996	MajGen Frank Libutti
June 1996–July 1998	MajGen John H. Admire
July 1998–May 1999	MajGen Michael W. Hagee
May 1999–July 2000	MajGen Gregory S. Newbold
July 2000–August 2002	MajGen James T. Conway
August 2002–August 2004	MajGen James N. Mattis
August 2004–August 2006	MajGen Richard F. Natonski
August 2006–May 2007	MajGen John M. Paxton Jr.
May–September 2007	BGen Richard P. Mills
September 2007–July 2009	MajGen Thomas D. Waldhauser
July 2009–August 2010	MajGen Richard P. Mills
August 2010–June 2011	MajGen Michael R. Regner
June 2011–June 2013	MajGen Ronald L. Bailey
June 2013–July 2015	MajGen Lawrence D. Nicholson
July–September 2015	BGen Daniel D. Yoo
September 2015–June 2017	MajGen Daniel J. O'Donohue
June 2017–present	BGen Eric M. Smith

FURTHER READING

Amero, Richard W. *Balboa Park and the 1915 Exposition*. Edited by Michael Kelly. Charleston, SC: History Press, 2013.

Barrow, J. M., and D. W. Porter. "Carriage House Stabilization at the Las Flores Adobe Site: Field School Training in the Conservation of Historic Adobe Structures." Paper presented at EarthBuild 2005, University of Technology, Sydney, Australia, 2005.

Barrow, J. M., E. L. Tolles, and S. J. Farneth. "Stabilization of the Landmark Las Flores Adobe: A Case Study of Restrained Seismic Retrofit." In *Structural Analysis of Historical Constructions*. Edited by Claudio Modena, Paulo B. Lourenço, and Pere Roca. London: A. A. Balkema Publishers, 2005.

Baumgartner, Jerome W. *Rancho Santa Margarita Remembered: An Oral History*. Santa Barbara, CA: Fithian Press, 1989.

Brigandi, Phil. "The Outposts of Mission San Luis Rey." *Journal of San Diego History* 45, no. 2 (Spring 1999): 1–5.

Camp Pendleton Archives, Museum Branch, MCIWEST-Marine Corps Base Camp Pendleton, San Diego, CA.

Camp Pendleton Scout. Weekly newspaper of Marine Corps Base Camp Pendleton.

Champie, Elmore A. *A Brief History of the Marine Corps Base and Recruit Depot: San Diego, California 1914–1962*. Revised ed. Marine Corps Historical Reference Series, No. 9. Washington, DC: Historical Branch, G-3 Division, Headquarters Marine Corps, 1962.

Chapin, 1stLt John C. *The 4th Marine Division in World War II*. Washington, DC: History and Museums Division, Headquarters Marine Corps, 1945, 1976.

Crawford, Danny J., Robert V. Aquilina, Ann A. Ferrante, and Shelia P. Gramblin. *The 1st Marine Division and Its Regiments*. Washington, DC: History and Museums Division, Headquarters Marine Corps, 1999.

Dunham, Maj George R., and Col David A. Quinlan. *U.S. Marines in Vietnam: The Bitter End, 1973–1975*. Washington, DC: History and Museums Division, Headquarters Marine Corps, 1990.

Engstrand, Iris H. W. "A Brief Sketch of San Diego's Military Presence: 1542–1945." *Journal of San Diego History* 60, no. 1–2 (Winter/Spring 2014): 1–26.

Geer, Andrew. *Reckless: Pride of the Marines*. New York: E. P. Dutton, 1955.

HDR, Inc. "Advanced Draft Report: Historic Resources Survey and Evaluation, Marine Corps Base, Camp Pendleton." Unpublished manuscript, July 2014.

Hebert, Edgar W. "Las Flores." *Journal of San Diego History* 7, no. 3 (July 1961): 31–35.

Photograph collection, Historical Reference Branch, Marine Corps History Division, Quantico, VA.

Hoffman, Nancy Lee White. "Sgt Reckless: Combat Veteran." *Leatherneck,* November 1992.

Hutton, Robin. *Sgt. Reckless: America's War Horse*. Washington, DC: Regnery History, 2014.

Jensen, James M. "John Forster: A California Ranchero." *California Historical Society Quarterly* 48, no. 1 (March 1969): 37–44.

Jordan, Ryan. "Remembering the Forgotten Village of San Onofre: An Untold Story of Race Relations." *Journal of San Diego History* 60, no. 1–2 (Winter/Spring 2014): 27–44.

Marine Corps Base Camp Pendleton: Celebrating 75 Years. Oceanside Chamber of Commerce in partnership with the Camp Pendleton Historical Society, 2017.

Mikesell, Stephen D., and Stephen R. Wee. *National Register Nomination: Santa Margarita Ranch House*. Washington, DC: National Parks Service, 1994.

Millett, Allan R. *Semper Fidelis: The History of the United States Marine Corps*. Revised ed. New York and Toronto: Free Press, 1991.

Nordhoff, Charles. *California: For Health, Pleasure, and Residence—A Book for Travelers and Settlers*. New York: Harper & Brothers, 1873.

O'Hara, Thomas. *Images of America: Camp Pendleton*. Charleston, SC: Arcadia Publishing, 2005.

Porter, Douglas, Steve Farneth, Sara Lardinois, Gee Heckscher, E. Leroy Tolles, Kelly Cobeen, Christian Carey, Julie Weisgerber, Liisa Reimann, and the Graduate Program in Historic Preservation at the University of Vermont. *Santa Margarita Ranch House Historic Structure Report*. San Diego, CA: U.S. Marine Corps and University of Vermont, 2007.

Santelli, Gabrielle M. Neufeld. *Marines in the Mexican War*. Edited by Charles R. Smith. Washington, DC: History and Museums Division, Headquarters Marine Corps, 1991.

Smith, Charles R., ed. *U.S. Marines in the Korean War*. Washington, DC: Marine Corps History Division, 2007.

The 3d Marine Division and Its Regiments. Washington, DC: History and Museums Division, Headquarters Marine Corps, 1983.

Thompson, P. L. "Rancho Santa Margarita." *Leatherneck,* September 1982.

Westermeyer, Paul. *U.S. Marines in the Gulf War, 1990–1991: Liberating Kuwait*. Quantico, VA: Marine Corps History Division, 2014.

Westermeyer, Paul W., and Alexander N. Hinman, ed. *Desert Voices: An Oral History Anthology of Marines in the Gulf War, 1990–1991*. Quantico, VA: Marine Corps History Division, 2016.

Witty, Robert M., and Neil Morgan. *Marines of the Margarita: The Story of Camp Pendleton and the Leathernecks Who Train on a Famous Rancho*. San Diego, CA: Frye and Smith, 1970.

Made in the USA
Monee, IL
05 November 2026

67385039R10063